DATE DUE

ILL 4/9/94 IL: 966771			
GAYLORD			PRINTED IN U.S.A.

The Courage to Survive

The Courage to Survive
The Life Career of
The Chronic Dialysis Patient

Mary Elizabeth O'Brien, R.N., Ph.D.
Associate Professor, School of Nursing
The Catholic University of America
Washington, D.C.

Grune & Stratton
A Subsidiary of Harcourt Brace Jovanovich, Publishers
New York   London
Paris   San Diego   San Francisco   São Paulo
Sydney   Tokyo   Toronto

Library of Congress Cataloging in Publication Data

O'Brien, Mary Elizabeth.
    The courage to survive.

    Includes bibliographical references and index.
    1. Hemodialysis—Psychological aspects.   2. Hemo-
dialysis—Social aspects.   3. Hemodialysis—Patients—
Psychology.   4. Adjustment (Psychology)   I. Title.
[DNLM: 1. Adaptation, Psychological.   2. Hemodialysis.
3. Kidney failure, Chronic—Rehabilitation.   4. Long
term care.   5. Patients.   WJ 378 013c]
RC901.7.H45027   1983   617'.461059   83-18612
ISBN 0-8089-1604-1

Grune & Stratton, Inc.
111 Fifth Avenue
New York, New York 10003

Distributed in the United Kingdom by
Grune & Stratton, Inc. (London) Ltd.
24/28 Oval Road, London NW 1

Library of Congress Catalog Number 83-18612
International Standard Book Number 0-8089-1604-1
Printed in the United States of America

This book is dedicated to the study group of hemodialysis patients, their families and friends, and their caregivers, with whom unique and treasured relationships have been established.

Contents

Preface   *xi*

1 The World of Maintenance Dialysis   *1*
  A Brief History of Dialysis in the United
    States   *2*
  The Study Problem: Long-term adaptation to
    Maintenance Dialysis   *4*
  The Research: The Life Career of the Chronic
    Hemodialysis Patient   *6*
  Overview   *9*

2 The Patients   *11*
  The Study Group   *12*
  Personality Changes   *14*
  Relationships with Significant Others   *26*
  Relationships with Caregivers   *29*
  Relationships with Other Dialysis
    Patients   *30*
  Dialysis Patient Friends' Deaths: "Pulling
    Back"   *32*
  Religion and Spirituality   *34*
  Modification of Life Goals   *38*
  Quality of Life   *39*
  Uncertainty of the Future   *42*

3 The Families   *48*
  Support of Significant Others   *49*
  Impact of ESRD and Dialysis on the
    Family   *55*
  Involvement with the Dialysis Treatment
    Regimen   *60*
  Relationships with the Dialysis Unit
    Staff   *62*
  Burnout   *62*
  Guilt   *65*
  Routinization of the Dialysis Regimen   *65*
  The Family Member as Potential Organ
    Donor   *66*

4   The Caregivers   *70*
        Characteristics of the Dialysis
        Caregiver   71
        The Role of the Physician   72
        The Role of the Nurse   73
        The Role of the Therapist/Technician   75
        The Role of the Social Worker   76
        Care Settings   77
        Staff Preparation, Experience, and
        Turnover   78
        "Elitism" of Staff   79
        Staff Support Systems   81
        Formal Staff–Patient Interaction   81
        The Chronic Patient Relationship: Staff
        Frustrations and Burnout   85
        Coping with Death   91
        A Caregiving Typology: The "Machine-tender,"
        the "Counselor," and the
        "Confidante"   95

5   The Dialysis Unit as a Social System   *101*
        Goals of the Dialysis Unit   102
        The Technical Environment   103
        The Physical Environment   104
        Territoriality—the Chair   105
        The Social Environment   106
        Staff–Patient Interaction   107
        The Waiting Room   108
        The Break Room   109
        Power—Authority and Decision-making in the
        Unit   110
        The Machine—Gift or Punishment   112
        The Rituals—"Going on" and "Coming
        off"   113
        The Language   115
        Stress in the Hemodialysis Unit   116
        Norms of Dialysis Patient Behavior   120
        Negative Sanctioning   122
        Ethical Dilemmas   123

6   Early Adaptation to Hemodialysis   *129*
        Stresses of Early Adaptation   129
        Early Regimen Compliance   133
        Evaluation by Dialysis Center
        Personnel   135
        Accomodation to a Hemodialysis
        Lifestyle   136

"Sickness–Wellness" Self-perception
Continuum   *140*
Secondary Gain   *144*

7   A Typology of Long-term Adaptation to
    Hemodialysis   *150*
    The Survivors   *151*
    Physical Changes Over Time   *153*
    Psychosocial Changes Over Time   *154*
    Support of Significant Others Over
    Time   *155*
    Long-term Compliance with the Treatment
    Regimen   *156*
    Compliance Behavior Changes Over
    Time   *157*
    Compliance Behavior and Sociodemographic
    Variables   *158*
    Survivors and Compliance Behavior   *159*
    A Typology of Adaptation to Long-term
    Hemodialysis: The "Career" Dialysis Patient,
    The "Part Time" Dialysis Patient, and The
    "Freelance" Dialysis Patient   *162*

8   Summary and Projections for the Future: The
    Courage to Survive   *170*
    Hemodialysis: The Long-term View   *170*
    Alternatives for the ESRD Patient   *174*
    Transplant Rejection: Return to the Dialysis
    Unit   *182*
    The Future of Dialysis: A Return to Scarce
    Resources?   *186*
    The Courage to Survive   *188*

    Appendix
    The Research Program: Evolution and
    Progress   *192*
    Study Purpose   *192*
    Program Phases   *192*
    Theoretical Framework   *193*
    Study Variables   *193*
    Operationalization of Variables   *194*
    Quantitative Method: Phase I, II, and
    III   *196*
    Qualitative Method: Phase III   *197*
    Observation: Phase III   *199*
    Intervention: Phase IV   *199*

    Index   *201*

Preface

The Courage to Survive is the result of a nine-year research effort during which the author attempted to identify and describe patterns of attitude and behavior adopted by long-term hemodialysis patients in coping with their illness and its regimen. Much has been written about the stresses of adaptation to dialysis, both for the patients and for their families. The distinction of the present work lies in the longitudinal nature of the study design, which ultimately provided the investigator with a panel group of patients who had experienced the hemodialysis regimen for an average of 9–12 years.

The themes of courage and survival emerged notably throughout the interviewing, not only with patients but with their significant others as well. The title *The Courage to Survive* is chosen deliberately, with full awareness of its similarity to the title of Renée Fox and Judith Swazey's classic work on organ transplantation and dialysis, *The Courage to Fail.* The first edition of *The Courage to Fail* (1974) was published just as the present research was being initiated, and provided impetus and support for continuing the study over time. Uncertainty, which was one of the predominant themes identified by Fox and Swazey, was found to be consistently present and to be of influence in the lives of the long-term dialysis patients involved in the research. This book is about their courage and their survival in the face of such overwhelming uncertainty.

The final phase of the research study, entitled "Effective Social Environment and Hemodialysis Adaptation," upon which major portions of this manuscript are based, was funded jointly by grants from Sigma Theta Tau (the National Honor Society of Nursing) and the Division of Nursing, Health Resources and Services Administration, U.S. Public Health Service, Department of Health and Human Services (Grant #R21 NU00824).

The author's deep gratitude is expressed to those who supported and encouraged this longitudinal research. Thanks go first to the dialysis patients, their families and friends, and their caregivers, who gave selflessly in the sharing of life experiences related to chronic renal failure and hemodialysis. Appreciation goes also to the dialysis unit physicians, administrators, and head nurses who facilitated entry to the patient study group. Numerous friends and colleagues assisted with the research, providing much invaluable assistance over the past nine years. These include Professors Hart M. Nelsen, John D. McCarthy, and Ann M. Douglas, who guided the initial phase of the study. Continuing consultation and support were provided by Professor Ada Jacox. Ms. Judith Strasser and Sr. Mariah Dietz assisted with data collection in the final phase of the research; and Dean Rosemary Donley contributed to the data analysis plan. Mrs. Mary Elizabeth Kentis was instrumental in the processing of much of the study's qualitative data; Dr. Rita Seifert assisted with the quantitative analysis.

Sincere thanks are due especially to Miss Julianne Mattimore, who carefully read and critiqued an early draft of the manuscript, and to Dr. R. Cletus Brady, who evaluated the entire work and has provided continuing consultation throughout the course of the research. And finally, deepest appreciation goes to Professor Renée C. Fox, whose concern and encouragement provided the author with the much-needed fortitude to complete the preparation of the manuscript.

——— 1 ———

The World of Maintenance Dialysis

*Waiting for tomorrow . . . asks for . . . a deep faith in the
value and meaning of life, and a strong hope which breaks
through the boundaries of death.*

Henri Nouwen
The Wounded Healer.

Hippocrates noted that "extreme remedies are very appropriate for extreme
diseases" *(Aphorisms)*. End-stage renal failure is an extreme disease. Dialysis
therapy is an extreme remedy. Renal dialysis, particularly the hemodialysis
regimen, in which one's continued survival depends upon sophisticated
technology, is perhaps the most extreme of the dialysis therapies. Thrice
weekly the patients must submit their beings to the "machine"—that magnificent, terrifying, beloved, despised mass of metal, wires and plastic, which
is capable of either preserving a life or taking a life, which can support daily
activities or hinder physical functioning, which may open the door to a
productive future or close that same door on a formerly bright career. In
the unique sociomedical environment of the hemodialysis unit, distinct patterns of interaction between patient and caregiver can be identified and
many ethical issues arise related to patients' survival and the quality of their
lives. Overt termination of a life may occur through a patient's failure to
participate in the dialysis procedure itself; covert termination of a life may
result from serious, continued noncompliance with the dietary regimen.
Patients, families, and caregivers struggle together with these and related
issues, trying to find the most acceptable, most humane, and least painful
resolution for all involved.

Within the past two decades, maintenance dialysis has become an accepted treatment modality for the patient with end-stage renal disease. In
1982, there were 466 free-standing and 654 hospital-based dialysis treatment

facilities operating throughout the United States[1] and the number continues to increase. With the advent in 1972 of Public Law 92-603, maintenance dialysis became available and accessible to all Medicaid and Medicare-eligible end-stage renal disease patients deemed to require such therapy.*

The numbers of new and long-term (10 years and longer) maintenance dialysis patients has greatly exceeded the early expectations of the health-care community. Approximately 68,000 Americans are currently being treated under the federally funded ESRD (end-stage renal disease) program at a cost estimated at approximately $1.2 billion for the 1982 fiscal year.[2] It is speculated that by 1985 the program will serve as many as 83,700 patients.[3]

A BRIEF HISTORY OF DIALYSIS IN THE UNITED STATES

Following the introduction of the first rotating drum artificial kidney by Holland's Willem Kolff in the early 1940s, several medical centers in the United States had by 1950 constructed their own hemodialysis machines. These centers included Georgetown University, Mount Sinai Hospital, and the Peter Bent Brigham Hospital.[4] In the mid-1950s, Kolff developed a disposable coil dialyzer and Fredrik Kiil of Norway introduced the flat plate parallel flow dialyzer for use in treatment procedures. At that time, however, the artificial kidney machine was utilized primarily for cases of acute renal failure, as access to an artery and vein could be obtained only by direct needle insertion or use of a surgical cut-down procedure. And, as Czackes and De-Nour note, "Obviously the number of times a suitable access to blood vessels could be achieved by these means was limited."[5] The problem was alleviated, however, by the development of the Quinton-Scribner external shunt in the late 1950s.[6] It is reported that the first chronic kidney patient was fitted with Belding Scribner's teflon-shunted cannulas in March of 1960; that patient survived on hemodialysis for 11 years.[7] Fox and Swazey observe that following the work of Scribner and his colleagues at University of Washington School of Medicine, "long-term hemodialysis became possible, making feasible the continuing treatment of patients with chronic, irreversible kidney failure."[8]

As more hemodialysis patients began using the external arteriovenous shunt, however, a considerable number of concerns about its adequacy were

---

*Medicare-eligible patients include those persons who meet stipulated medicare criteria by virtue of age and work history and who are certified ESRD patients. The program covers 80 percent of all illness-related costs. The remaining 20 percent is generally covered by private health insurance, or by Medicaid for the medically indigent patient.

identified. These included such issues as patency, or the maintaining of an adequate blood flow through the teflon cannulas; dangers, such as possible infection or the accidental disconnection of one or both cannulas; and the unpleasant cosmetic effect (patients were required to cover the AV shunt with unsightly bandages). In recent years progress has been made in vascular access with the introduction of the surgically-created internal arteriovenous fistula, and several types of graft AV fistulae. One of the most commonly used prosthetic grafts is the polytetrafluorethylene or "Gore-Tex" graft.

— In-center hemodialysis was first introduced in the form of hospital-based outpatient dialysis units during the mid to late 1960s, with smaller facilities of approximately 10 to 15 patient stations being the norm. During the late 1960s and early 1970s, larger freestanding outpatient centers began to open. With the passage of Public Law 92-603 in 1972, the growth of all in-center dialysis facilities escalated significantly to accommodate the rapidly growing hemodialysis patient population.

Home hemodialysis began around 1963.[9] It is reported that close to 40 percent of the dialysis patient population had initiated home dialysis in 1972; however, by 1978 the home dialysis group had decreased to 12–16 percent of the total patient population.[10] Currently the home dialysis population seems to have stabilized at around 17 percent of the national dialysis patient group.[1]

Intermittent peritoneal dialysis (IPD) was first introduced as a treatment for renal failure in 1927, but was not widely accepted until the 1960s.[11] Initially, IPD was used in a limited manner to treat patients with acute renal failure or those who are awaiting hemodialysis access or kidney transplantation. The fairly recent advent of automatic cycling machines,[12] however, and of a permanently implantable silicone catheter, greatly enhanced the potential of IPD as a treatment modality for end-stage renal failure.

Presently, peritoneal dialysis, in two of its newer modalities, continuous ambulatory peritoneal dialysis (CAPD) and continuous cyclic peritoneal dialysis (CCPD) is being advocated for a number of dialysis patients. CAPD is perhaps the most widely accepted peritoneal dialysis modality currently in use. CAPD was introduced in the mid-1970s and has been acclaimed by some as "the treatment of choice" for a select group of ESRD patients. It has been claimed that CAPD "may cause fewer symptoms such as headaches, cramps, hypertension, and fatigue as compared with hemodialysis."[13] Generally, fewer dietary and fluid restrictions are necessary, also. There remain some concerns about the procedure, relating to such problems as peritonitis and patient noncompliance, but it has been estimated that once these issues are resolved, CAPD could be the modality of choice for approximately 10–35 percent of the ESRD population.[14]

CCPD is another recently advocated home dialysis modality. It involves an overnight peritoneal dialysis procedure theoretically to be carried out while the patient sleeps. It is suggested that this procedure works well for the patient who must be unrestricted during the daytime hours. A patient can learn to operate the automatic cycler quite easily and may dialyze alone without fear. One manufacturer of the CCPD equipment labeled his cycling machine the "R2D2" of dialysis technology. (In-center hemodialysis, home hemodialysis, IPD, CAPD, and CCPD are discussed individually in Chapter 8.)

THE STUDY PROBLEM: LONG-TERM ADAPTATION TO MAINTENANCE DIALYSIS

Kidney disease is reported to be "the fourth most frequent cause of death in the United States today,"[15] yet in many persons the condition may lie dormant for years before any serious manifestations occur. At first, the warning signs of a kidney problem may be passed over or attributed to some other cause. These symptoms, which include such occurrences as headaches, nausea, and changes in blood pressure, may also be transient, and unless the disease is documented by laboratory studies and clinical examination, the condition may be totally ignored until it is rapidly approaching the critical stage of almost complete kidney shutdown and uremia. The suddenness of the onset and the rapidity with which ESRD progresses to a critical state is usually a very serious problem both for patients and their families in that they have had no warning; on the point, Rapaport explains: "The fact that this acute, life-threatening emergency can often develop at a time when neither the patient nor his family is prepared to cope with the consequences is a particularly striking feature of this form of kidney disease."[15]

Although the causes of end-stage renal failure are numerous (including such conditions as glomerulopathies, hypertensive nephrosclerosis, polycystic kidney disease, pyelonephritis, and diabetic nephropathy), three common stages, which may overlap to some degree, are usually identified. The phases, as presented by Harrington and Brener,[16] include (1) diminished renal reserve, (2) renal insufficiency, and (3) uremia. During the first two of these stages renal function is somewhat impaired and metabolic wastes begin to accumulate in the blood, but symptoms may be mild to moderate. During the stage of uremia, however, urine output usually decreases sharply, fluid and electrolyte balance is disturbed, and numerous debilitating symptoms occur. It is as the patient progresses toward the onset of this stage that the consideration of hemodialysis or other modality of treatment is entertained. As Pendras and Erickson explain: "It is extremely important to start the chronic uremia patient on hemodialysis as soon as symptomatic uremia makes it impossible for him to discharge his duties at work and at home."[18] These authors hold that early and aggressive use of hemodialysis treatment for chronic renal patients greatly enhances the patient's ability to adjust to and accept the total dialysis regimen.[17]

Holcomb found that chronic renal disease and hemodialysis impose great stress on both the patient and family,[18] and the American Association of Nephrology Social Workers suggests that social-psychological factors are involved in the adjustment of all chronic renal patients who must not only face the possibility of death but must "select and incorporate a machine or a new kidney into their system of self as a basis for living; and find meaningful roles within the family and society."[19] In a study of 14 hemodialysis patients at a university medical center, it was reported that all respondents either overtly or covertly described personal feelings of fear of death: "This fear was their immediate, initial and recurring reaction to learning of the diagnosis of chronic and progressive renal failure."[20]

Once the treatment procedure has been initiated, most hemodialysis patients are placed on a regimen requiring approximately 3–4 hours on the machine, two or three times each week, and adherence to certain dietary and fluid restrictions. Obviously such a regimen will necessitate certain self-threatening changes and modifications of social interactions and activities in the areas of family life, work, and recreation. According to Glassman and Siegel, "The necessity for frequent hemodialysis has caused a marked change in the life-style and mobility of all patients."[21] Goldberg reports that, in regard to the area of dialysis patients' interactions, "opportunities to participate in social and community activities have been reduced."[22]

As noted above, studies have shown that persons undergoing dialysis treatments on a consistent and ongoing basis must cope with numerous physical and psychosocial stresses. Due to the relatively brief history of the therapeutic modality, however, minimal data are available regarding "long-term adaptation" to the therapeutic regimen. Even the phrase "long-term adaptation" changes in definition as patients survive for longer periods of time on maintenance dialysis.

Adaptation to any chronic illness is difficult and may be quite frightening. Both patient and family members are required to initiate new or previously unused coping behaviors, and adjustments must frequently be made in the allocation of usual responsibilities. In some cases family roles are seriously disrupted. The chronically ill person may also need to reorient his or her self-concept in order to incorporate a "sick" or "disabled" status into the usual repertoire of attitudes and behaviors.

For the maintenance dialysis patient it is suggested that many other unique sources of social-psychological stress can be identified; the specific details of these stresses, however, are somewhat elusive and difficult to define. Czaczkes and De-Nour assert that while much has been written about psychological stress for the chronic dialysis patient, what is needed is a greater understanding of the sources of such stress. They note: "If we know the sources of stress, we might differentiate between those inherent in the procedure and those that could be modified, thus reducing the stressfulness of the situation."[5] One factor that has frequently been associated with the concepts of stress and adaptation for the maintenance dialysis patient is that of support or lack of support from significant others. This variable has important implications for adaptation over the long run. A dialysis patient is often surrounded by friends and family members during the early period of renal failure, when the dialysis procedure is being initiated. As the "novelty" of the condition diminishes, however, so, frequently, does the interest and concern of others. Support of a chronically ill patient may wear extremely thin over time. It is suggested then that the stresses on a patient's family or significant others need also to be examined and addressed in the course of therapy for maintenance dialysis.

Thus, the long-term dialysis patient, the patient who has coped with ESRD and dialytic therapy for a number of years, has been, and is continually, faced with numerous and diverse threats to physical and emotional well-being. What then, one must ask, have been the life experiences of such patients? What have been the related experiences of their significant others? What have been the coping behaviors adopted by their caregivers? And how do the interactions and adaptive mechanisms of all three groups mesh to form the tapestry in which is woven the life career of the maintenance dialysis patient?

THE RESEARCH: THE LIFE CAREER OF THE CHRONIC HEMODIALYSIS PATIENT

The longitudinal research program upon which this book is based was begun in 1974. At the time of the study's initiation, a growing number of patients were being hospitalized for access (shunt or fistula) surgery, preparatory to commencing hemodialysis therapy. These patients frequently expressed their fears, anxieties, and general concerns about how difficult it was and would be in the future to live with end-stage renal failure and the dialysis regimen. Such patient comments led the investigator to wonder how dialysis patients felt about particular concerns, how they coped with their condition, and what could be done to support and facilitate adaptation to this life-threatening illness and its complex treatment modality. Thus, out of clinical experience with a number of ESRD patients, the research program, beginning with a correlational study relating to the psychosocial adjustment of the chronic dialysis patient, was born. As the research progressed over a 9-year period, both quantitative (structured interview) and qualitative (unstructured interview and discussion; observation) methods were employed to examine the life career of long-term hemodialysis patients. The subjects consisted of a panel group of dialysis patients, their self-identified significant others (family members and friends), and their caregivers. The study topics examined included the relationships between social support and social and psychosocial adaptation for the dialysis patient; change over time in selected variables, including social support, interactional behavior, quality of interaction, sick role behavior (compliance with the treatment regimen), secondary gain, and alienation; the pattern or the multiple patterns of patient attitudes and behaviors involved in adapting to long-term dialysis; the family or friendship group dynamics associated with provision of support for a maintenance dialysis patient; the mode of coping adopted by dialysis unit personnel involved in caregiving relationships with long-term patients; and finally, the characteristics, attitudes, and behaviors evidenced in the treatment setting (the dialysis unit) viewed as a social system.

The patient group studied was comprised of adult hemodialysis patients who were receiving treatment at either hospital-based or free-standing outpatient dialysis units in an East Coast metropolitan area. During the original sample selection, any patient with serious complicating physical or psychological conditions other than end-stage renal failure was excluded from the study. At the time of the initial phase of the research, Time 1, 126 patients were interviewed utilizing a structured interview schedule; three years later, Time 2, a panel group of 63 (50 percent) of the original patients were located and reinterviewed (of the other 63 patients in the original sample, 7 were lost through physical migration, 1 had switched to home dialysis, 1 was attending treatment sessions irregularly and he subsequently died, 7 had been successfully transplanted, and 47 were deceased). The group of subjects interviewed at Time 1 was found to be representative of the total dialysis population in the United States in regard to both age and sex.[23] Of the panel group reinterviewed at Time 2, approximately half were male and half female; 32 patients (approximately 50 percent) were married; and ages were widely distributed between 21 and 70, with the greatest number of respondents falling within the age category of 30–59 years.

After another 3-year interval, Time 3, a new panel group, constituted of 33 of the 63 patients interviewed at Time 2, were again interviewed utilizing the originally developed instruments. (All 30 of the T2 panel patients not interviewed at T3 had died). At Time 3, new methodologies and new data sources were introduced into the research. New qualitative methodologies were adopted, in that now both focused unstructured interviewing and discussion were carried out with the dialysis patients, as well as the formal structured interviews. Patients were requested to articulate their feelings about many aspects of adjustment to long-term hemodialysis, and whenever possible these interview sessions were recorded. New data sources were introduced at Time 3; each patient was asked to identify a significant other (either family member or close friend) and the patient's permission was requested to contact the named individual to seek an interview. The population of significant others consisted of 26 study respondents, including 12 spouses of dialysis patients, 4 friends, 3 mothers, 1 cousin, and 6 sons or daughters. Finally, a group of 45 long-term hemodialysis unit caregivers were interviewed regarding their perceptions of patient adaptation and, more importantly, their own attitudes and the usual behaviors involved in coping with the care of the maintenance dialysis patient. The caregiver group that was interviewed consisted primarily of dialysis unit head nurses, staff nurses, and therapists. Data were also elicited through interviews with several physicians, social workers, machine technicians, and clerks associated with chronic dialysis units. Most of the study dialysis unit personnel had been working in the field for five or more years at the time of interview.

Focused interview guides were developed for conversations with patients, caregivers, and family members in order to direct the questioning toward certain aspects of long-term dialysis adaptation. However, respondents were also encouraged to express freely any of their own thoughts, problems, or concerns that the interviewer did not identify, in hope of eliciting serendipitous themes or patterns of attitudes and behavior.

In order to construct as complete a picture of the life career of the dialysis patient as possible, one other dimension was added to the study at Time 3—an evaluation of the dialysis unit as a social system. To identify relevant characteristics, attitudes, and behaviors of actors in the setting, direct observation was carried out in several large metropolitan hemodialysis units. An observation guide was developed to sensitize the nonparticipant observers to appropriate data sources.

During the course of the 9-year research program, the investigator came to know a number of the study respondents quite well. In many cases homes were visited in order to contact patients or their significant others, and some close and valued relationships developed through the data collection process. Often study subjects shared their positive reactions to the research. The wife of a long-term dialysis patient commented that it was "really good" to have someone to talk to about her husband's condition. And a caregiver expressed satisfaction that she was being approached for an interview. She observed, "A lot of people interview dialysis patients. It's about time somebody talked to the staff and asked us how we feel."

Through the evolution of the study many patient changes were observed, both positive and negative. Some patients improved notably over time, initiating a greater degree of selfcare and more positive attitudes toward the quality of their lives; adaptation to the complex dialysis treatment regimen was accomplished and routinized in order to avoid conflict with work and family responsibilities. Unfortunately, physical deterioration occurred in other study subjects, all too often resulting in the patient's death. Frequently, a friend or family member called the researcher after such a death.

As the spouse of a long-term patient commented, "I knew that you would want to know." As the research program continues to evolve, further change in study patients will be carefully charted. The methodological details of the study are presented in the appendix.

OVERVIEW

Through the research process, a measure of new knowledge and understanding of adaptation to maintenance dialysis has been gleaned. Patients, family members and friends, and caregivers openly and generously shared their concerns, their sufferings, and their joys in the hope of helping others who may be destined to tread a similar path. These concerns, these sufferings, and these joys are described in the pages which follow. Whenever appropriate, comments are presented in the study respondents' own words.

Chapter 2 consists of an examination of issues involving the hemodialysis patients in their long-term adaptive and survival mechanisms. A description of the family and friendship group dynamics involved in the support over time of the maintenance dialysis patient is presented in Chapter 3. Chapter 4 examines the coping behaviors of dialysis unit personnel in long-term caregiving relationships with ESRD patients. An analysis of the hemodialysis unit as a unique sociomedical environment, where many norms of the patient–caregiver relationship may be modified, is presented in Chapter 5. Chapter 6 focuses upon early adaptation to the treatment regimen and presents rationale for a dialysis patient "sickness–wellness self-perception continuum." In Chapter 7, a typology of patient attitudes and behaviors associated with long-term hemodialysis adaptation is described. Three adaptive types are discussed: (1) the "career dialysis patient," (2) the "part-time dialysis patient," and (3) the "freelance dialysis patient." Chapter 8 contains a summary and projections for the future.

REFERENCES

1. Robinson D: Kidney dialysis: A taxpayers' nightmare. Readers Digest, October: 149–152, 1982, p 150
2. Capelli JP: Testimony of the Catholic Health Association of the United States on proposed end-stage renal disease regulations, before the Subcommittee on Oversight, Committee on Ways and Means, United States House of Representatives, April 22, 1982
3. Greenspan RE: The high price of federally regulated hemodialysis. JAMA 246: 1909–1911, 1981, p 1909
4. Fox RC, and Swazey JP: The Courage to Fail (2nd ed). Chicago, The University of Chicago Press, 1978, p 201

5. Czaczkes, JW, De-Nour AK: Chronic hemodialysis as a way of life. New York, Brunner/Mazel, 1978, p 22
6. Lancaster LE: The patient receiving hemodialysis, in Lancaster LE (Ed): The Patient with End-Stage Renal Disease. New York, Wiley, 1979, pp 140–181
7. Katz J, Capron AM: Catastrophic Diseases: Who Decides What? New York: Russell Sage, 1975, p 37
8. Fox RC, Swazey JP: The Courage to Fail (2nd ed). Chicago, the University of Chicago Press, 1978, p 202
9. Gutch CF, Stoner MH: Review of Hemodialysis for Nurses and Dialysis Personnel (2nd ed). St. Louis, C. V. Mosby, 1975
10. Wineman RJ: End-stage renal disease. Dialysis and Transplantation 7:1034–1038, 1978
11. Schmidt RW, Blumenkrantz MJ: IPD, CAPD, CCPD, CRPD-peritoneal dialysis: Past, present and future. Int J Artif Org 4:124–129, 1981, p 124
12. Tenkhoff H, Meston B, Shilipetar G: A simplified automatic peritoneal dialysis system. Trans Am Soc Artif Intern Organs 18:436–440, 1972
13. Schmidt RW, Blumenkrantz MJ: IPD, CAPD, CCPD, CRPD-peritoneal dialysis: Past, present and future. Int J Artif Org 4:124–129, 1981, p 125
14. Schmidt RW, Blumenkrantz MJ: IPD, CAPD, CCPD, CRPD-peritoneal dialysis: Past, present and future. Int J Artif Org 4:124–129, 1981, p 127
15. Rapaport FT (Ed): A Second Look at Life. New York, Grune and Stratton, 1973, p 1
16. Harrington JD, Brener E: Patient Care in Renal Failure. Philadelphia, W.B. Saunders, 1973, p 46
17. Pendras JP, Erickson RJ: Hemodialysis: A successful therapy for chronic uremia. Arch Intern Med 54:293–311, 1966, pp 306–307
18. Holcomb JL: Social functioning of artificial kidney patients. Soc Sci Med 7:109–119, 1973, p 109
19. Kari J, Kyle E, Vivalda E: Positions of the nephrology social workers. Dialysis and Transplantation 3:12–17, 1974, p 12
20. Beard, BH: Fear of life and fear of death. Arch Gen Psychiatry 21:373–380, 1969, p 379
21. Glassman B, Siegel A: Personality correlates of survival in a long-term hemodialysis program. Arch Gen Psychiatry 22:566–574, 1970, p 573
22. Goldberg RT: Vocational rehabilitation of patients on long-term hemodialysis. Arch Phys Med Rehabil 55:60–65, 1974, p 64
23. Statistical Analysis of Patients in Reporting U.S. Dialysis Centers. Paper prepared for the Artificial Kidney Chronic Uremia Program, N.I.H., Bethesda, Maryland. Chapel Hill, North Carolina, Research Triangle Institute, 1975

2

The Patients

You need to know what kind of person has the disease, not what kind of disease the person has.

Sir William Osler

End-stage renal disease is decidedly not a selective illness. The hemodialysis patient population consists of a group ranging from toddlers to the elderly, from the very poor to the very wealthy, and from the highly educated to the illiterate. While there appears to be a preponderance of patients in the middle to older age ranges relative to specific disease entities such as the adult form of polycystic kidneys, overall ages of ESRD patients vary notably. Although "the early stages of renal disease are quite variable, the end-stages of renal disease are surprisingly similar, and in many cases the original cause cannot be identified".[1] The possible choices of treatment for end-stage renal disease, therefore, are consistent for most patients: hemodialysis (in-center or home), peritoneal dialysis (IPD, CAPD, or CCPD), and kidney transplantation.

It has been noted by Lancaster and Pierce that end-stage renal failure so affects normal body functions as to alter significantly a person's quality of life: "Hardly an aspect of physical, social or psychological performance is left untouched by this disease process."[2] Czaczkes and Kaplan De-Nour comment that the treatment modality of chronic dialysis is "stressful for everybody involved—for the patient, for the family and for the staff."[3] Numerous discussions on both the physical and psychosocial adaptation of the chronic hemodialysis patient have been presented in the literature; the present chapter, however, addresses patient adjustment and its correlates from the perspective of longitudinal research. Frequently examples of such adaptation-related factors as stressors, coping behaviors and the import of support systems are presented in the patient's (or significant others') own words.

THE STUDY GROUP

In the initial phase of the study a group of 126 chronic hemodialysis patients was identified and interviewed (Time 1, 1974–1975) These patients were, at that time, found to be generally representative for age and sex of the reported national adult dialysis population as presented in 1975.[4] The study population showed a curvilinear distribution of patients in regard to age, with the highest number (39) or 31 percent of the group falling into the 50–59-year-old category. Ages ranged from 21 to 70. Seventy-five, or 59.5 percent of the study respondents were males, and 82 (65.1 percent) were black. The racial composition of the study group was assumed to be unrepresentative, as the study had been undertaken in an urban area with a disproportionately large black subpopulation. The sample group was equalized to the extent possible, however, through the inclusion of one partially suburban facility.[5] Ninety-two (73 percent of the group) had a high school diploma or less. Thirty-four (27 percent) of the patients had some college or graduate school experience. Of the total study sample, 83 respondents (65.9 percent) fell into the occupational categories of "unskilled or semi-skilled"; 28 (22.2 percent) were categorized as "professionals."

In the course of the study it was found that a high percentage of the patients, approximately 50–75 percent, were unemployed or able to work only infrequently. A partial explanation of this finding might be related to the high percentage of respondents who reported their occupation to fall within the categories of unskilled or semi-skilled. Patients with renal failure who are undergoing dialysis treatment theoretically should be able to function fairly normally in regard to routine daily activities, including work. However, if their work is of such a nature as to require great expenditures of physical energy, such as that often demanded of unskilled or semi-skilled manual occupations, the patient is frequently too physically fatigued to handle a regular routine. It is interesting to note that of those persons who responded positively to carrying out a full-time job, the majority were professionals, e.g., pharmacist, physician, librarian, corporation executive, government analyst, and the like. Most women respondents reported that they were able to take care of their own family and housework responsibilities, with the exception of very heavy duties, which they relegated to others.[6]

There was a wide variety of income levels among the dialysis patients, with the modal grouping of 33 (26.2 percent) falling into the $3,000 to $4,000 category. Possibly this high percentage of patients in a low-income group was due to the number of respondents who were, because of their illness, unable to work and had resorted to public assistance. At the time of original interview, 63 (50) percent of the study respondents were married. Seventeen patients (13.5 percent) categorized themselves as never having been married and the remaining 46 (36.5 percent) fell into the groups of widowed, divorced or separated. The largest percentage of subjects, 57 (45.2 percent) reported living with other adults, while 38 (30.2 percent) lived with both other adults and children. Only one male reported living with children only and 17 males (13.5 percent) lived alone.

In regard to religious affiliation, 87 (69.1 percent) of the patients identified themselves as Protestant; 25 (17.8 percent) Catholic, 8 (6.3 percent) Jewish, and 6 (4.8 percent) acknowledged no formalized religious belief system. One respondent suggested an agnostic position by noting that he lived as though there might or might not be a God and attempted to follow his own personal moral code.[7] Data reflective of hemodialysis history revealed that the group profile was as follows: 34 (27 percent) of the patients had initiated dialysis during the preceding year; 27 (21.4 percent) had initiated dialysis 1–2 years ago; 32 (25.4 percent), 2–3 years ago; 18 (14.3 percent), 3–4 years ago; 11 (8.7 percent), 4–5 years ago; and 4 (3.2 percent) patients reported initiating dialysis five or more years previous to the initial interview. The average length of time on the machine was 2.2 years for the total group; no patient reported receiving hemodialysis for more than six years.

Three years after the initial data collection period (1977–1978), Time 2, 63 patients of the original study group were located and re-interviewed. The newly established panel group was almost equally divided between men and women; ages were distributed between 24 and 69; 25.4 percent were white and 74.6 percent were black; and 11 of the patients had a hemodialysis treatment history of approximately 8 or more years. In regard to religious affiliation, 50 of the subjects were Protestant; 10, Catholic; 3, Jewish; and none reported having no religious group identification at this time.

Of the 63 patients not interviewed at Time 2, 47 had died; 7 had been successfully transplanted; one had transferred to home dialysis; one had developed a pattern of unstable hemodialysis treatment session attendance; and 7 were lost through physical migration (though still on dialysis in other geographical areas). In reviewing records of the patients who had died, the following causes of death were identified: renal complications—20; cardiac arrest—6; stroke—2; failure to thrive—3; posttransplant complications—4; congestive heart failure—2; myocardial infarction—2; brain lesion—1; cerebral dysfunction—1; accidents (car accident, shooting)—2; and unknown cause home deaths—4.

After another three-year period (1980–1981), Time 3, members of the hemodialysis patient panel group were again contacted. Thirty-three patients were located and re-interviewed; the other 30 patients of the panel group had died during the period between study Time 2 and Time 3. In addition to carrying out fresh structured patient interviews, several new elements were incorporated into the study at Time 3. These included a plan to visit the patient again, either at home or in the unit, to conduct informal focused interviews in order to identify particular problems and stresses involved with long-term adaptation to maintenance hemodialysis; a request to interview one of each patient's significant others—either a family member or friend; and interviews with a sample group of professional and paraprofessional hemodialysis unit staff members. (Methodological details of the research are presented in the Appendix.)

Through the above interviews, data were elicited relative to the long-term adaptation and life career of the maintenance dialysis patient. This chapter presents patient, family or friend, and caregiver's comments and reflections in regard to the long-term hemodialysis patient's career, focusing on such factors as personality change (depression, alienation, dependency); stigma; social functioning (family life, work, recreation); sexual functioning; relationships with significant others; religion and spirituality; modification of life goals; quality of life; and uncertainty of the future.

PERSONALITY CHANGES

It has been asserted that, "Chronic renal failure and its treatment modality probably disrupt and reorganize more aspects of a patient's life than do other chronic illnesses."[8] Kaplan De-Nour et al reported that, in evaluating the emotional reactions of hemodialysis patients, "Surprising changes in combination and intensity of defenses over short periods of time were observed, resulting in ever-changing clinical pictures, changing behaviors and facets of personality."[9] The American Association of Nephrology Social Workers highlighted the uniqueness of the psychosocial adjustment necessitated by end-stage renal disease in pointing out that patients not only face the possibility of death, but must also "select and incorporate a machine or a new kidney into their system of self as a basis for living; and find meaningful roles within the family and society."[10]

The wife of a long-term dialysis patient in the present study described the impact of her spouse's personality modification this way:

I was ready to adapt and adjust our lives if necessary but I was not expecting what happened. I was not expecting the personality change. For him to become as demanding, I mean really demanding, as he is. He will not ask anybody else to do anything for him except me. He really had mood swings. Sometimes he's up and everything's fine and other times he really gets irritable. When I learned that the disease caused it, I could accept more of the personality changes and things that go with it. He's very definite in a lot of things that he thinks he remembers, but he gets them mixed up—but he doesn't think so, and he gets upset if I correct him.

A number of significant others, especially patient spouses, echoed similar complaints. Usually their concern, as well as that of the patients, focused upon one of the following maladaptive responses: depression, alienation, or dependency. The concept of stigma was also identified as a negative situational response evoked by the patient's life-threatening illness and the uniqueness of the associated treatment.

Depression

Czaczkes and Kaplan De-Nour, after reviewing a number of published reports, concluded that the evidence in the extant literature "indicates that the majority of patients on chronic dialysis are depressed."[10] One of the notable problems articulated by patients and their families and caregivers in the present study was that of depression, or mood change, as it was sometimes described. Patient respondents expressed their concept of depression variously as "feeling low about yourself and about the future"; getting "fed up and just not caring much about anything any more"; being "really down in the dumps about this thing [ESRD and dialysis]"; and one patient complained: "Sometimes you just don't feel like you can go on you're so low—you just don't think you can make it in to this place [the unit] one more time." Patients' family members and caregivers commented frequently on depression. As the wife of a hemodialysis patient put it, "Sometimes A. gets very depressed and then he's irritable. It was difficult at first but now I understand it's from his disease and I can kind of ignore it. If he snaps, I can say, 'What's the matter with you?' and pass it off." Another spouse added, "His [her husband's] moods change real often and then he gets down low and it's hard—sometimes I'm afraid to leave the house—to leave him alone at those times, but then, too, he'll snap out of it." A dialysis unit caregiver summed it up this way: "These patients do get depressed—they have a right to. It's [dialysis and ESRD] a lot to face." But she added, "They usually bounce back. We try to help and some of their families do too, and they [patients] usually come out of it, but you know that it [depression] is probably going to hit again, so you might as well be ready."

Alienation

Alienation among hemodialysis patients was examined quantitatively in the present research to determine change over time as well as relationship to the support of significant others; this is discussed in detail in Chapter 7.

Alienation, as self-perceived by dialysis patients, however, was frequently referenced in focused interviews under the label of "loneliness."

Regarding social alienation, Schlesinger holds that any disability or long-term illness may result in impaired communication, problems in regard to social goal-setting, and inadequacy in the assumption of appropriate role behaviors.[11] He also suggests that various factors hinder the patient in the process of social reintegration; for example, the factor of social-role-responsibilities, duties, and life goals may have changed; family changes may also have occurred, members having assumed different decision-making responsibilities and tasks; and friends may have developed new interests. The patient condition may make reestablishment of the status quo difficult as the behavior of others toward one may have changed significantly, and the change is not usually consistent.[12] Sorensen reinforces this suggestion: "Although dialysis patients are usually surrounded by friends early in their course, the novelty soon wears off, and they tend to become socially isolated. It is well known that our society tends to draw away from chronically ill patients."[13]

For hemodialysis patients who live constantly facing the possibilty of death, alienation may also come about because of their own depression and personal withdrawal from former friends and activities. Nordan et al. assert that the extreme psychological stress that occurs in relation to the patient, as well as in relation to the patient's family and society, "often leads to depression so severe as to become incapacitating, ending in complete withdrawal into the self.[14]

One dialysis patient in the present study described his situation of alienation or "loneliness" as follows: "I get real lonely. I hardly have any friends any more. My blood [chemistries] is okay, but I'm tired all the time and I don't have the time or feel like goin' out." Another patient put it this way: "Well, you get isolated from people [because] you have to go on dialysis. If you socialize with a group of people and the time that you have to go on dialysis may be the time that all of them are going out, it gets to be a problem. Or something could happen and you might be sick and you can't participate in their activities. So you just have to adjust yourself." The subject of former activities was often linked to alienation. For example, one patient said, "I'm alone a lot. I can't run the streets like I used to. So, some of my old drinking buddies don't hang around much any more. You find out about real friends when there's trouble." Another respondent stated that what she missed most since she couldn't do so much physically was "doin' what I used to do— my friends and my people comin' around." A female patient reported that she had several siblings living in the area who used to visit her, but added ruefully, "I didn't die fast enough for them, so they stopped coming to see me at all." The comments of two long-term hemodialysis patients might be said to encapsulate the group's self-perceived alienation. A woman patient noted, "It's lonely on the machine—people feel sorry for you but they don't really understand"; and a male respondent simply said, "You have to be on that machine to really know." Not all hemodialysis patients, however, perceived themselves as alienated. A number reported that their lives were very full and active and that they were in fact able to engage in and maintain normal family and friendship group relationships. These patients generally admitted to little or no depression and/or dependency as well. No demographic correlates of alienation, such as age, sex, race, socioeconomic status, or marital status appeared consistent or relevant.

Dependency

The concept of dependency—depending upon family, friends, caregivers, technology, and even society—is frequently identified in discussions of the psychosocial adaptation to maintenance hemodialysis. Kaplan De-Nour and Czaczkes comment, "The life of the dialysis patient is not only machine-dependent but also man-dependent, starting from the man [men] who has selected [accepted] the patient for treatment, going on to the team that operates [controls] the machines, and often also to the society that pays for the treatment."[15] Dependency is related by Brundage to the dialysis patient's physical limitations and need for assistance with the treatment procedure; she focuses on the struggle of dependence versus independence, pointing out that, while the patient may be forced into dependency on the artificial kidney, he or she may strongly resent this imposition and become very difficult to deal with; she writes: "Reaction to dependency may include aggressive behavior [displacement], acceptance, leading to extreme dependence and anxiety.[16] It has also been reported that patient dependency upon spouse and the spouse's stress in regard to dialysis are primarily related to previous dependency needs, or "dependency vs. dominance within the family."[17]

A considerable number of patient subjects in the present research openly admitted to their dependency, especially upon family members or friends. One patient stated, "I'm really dependent upon my daughter to help me and it's hard. She has her own children and her own life to live, but she has to do for me, too. I can't do for myself now." Another commented, "It's really tough to have to have other people drive you, do for you. I used to do whatever I wanted and go wherever I wanted, but I'm just not up to it anymore." Obviously, dependency and dependency needs varied with changes in the patient's physical condition. As a male dialysis patient described it, "Right now I'm totally dependent upon my wife in every way possible. Up until last year I was doing everything by myself. When I left my job, I worked part-time, but my legs gave out. It's really difficult having to be dependent like this." Caregivers commented on patient's dependency in the dialysis unit, suggesting also that the degree of dependency frequently was associated with the patient's physical needs.

It was also found that not all patients exhibited dependency needs or behavior to a similar degree, even in the case of severe physical incapacity. One dialysis unit staff nurse commented, "I think there are a certain number of patients who have a dependent personality and who are manipulative in other ways. They take advantage of being ill for a secondary gain. But I don't think that's true of all patients. And I think, also, that in lots of cases it's something that can be worked with." In some situations, family members, particularly spouses, were viewed by caregivers as encouraging dependency on the part of the patient; this was reportedly associated with the family member's own needs or feelings of guilt in relation to the illness condition.

Stigma

Erving Goffman noted that the concept of "stigma" was originated by the Greeks, who used the term to refer to certain bodily signs that had a negative connotation regarding the moral status of an individual. Goffman added that, "today the term is widely used in something like the original sense, but is applied more to the disgrace itself than to the bodily evidence of it.[18] Three types of stigma are identified, the first of which is described as related to abominations of the body—the physical deformities.[18]

Although the stigma concept has not often been formally associated with end-stage renal failure and dialysis, it sometimes surfaces subtly in referencing the uniqueness or differentness of the situation. Hansen highlights the stress caused for ESRD patients by the loss of urination and notes, "The loss of this very normal function only further emphasizes to the patient the fact that he is different. When the nonfunctioning kidneys are actually removed in preparation for transplant, or perhaps to control blood pressure, the whole finality of his disease becomes a reality to him."[19] And Dickerson, in discussing the stresses of hemodialysis, quotes a patient as saying, "If my water stops, I'll just die ... I won't be a man any longer and there will be nothing to live for."[20]

Many study respondents referred to the conceptual undertones of stigma, although the term was not directly applied. One female patient described it this way: "Kidney failure can be embarrassing. It's strange not to be able to pass urine. I used to say I had to run to the ladies room so people wouldn't think it was funny that I went all day without going to the bathroom." She added, "I don't perspire either—it's really hard in the summer. When I feel a drop of sweat on my face I get all excited because then I feel normal." A male patient commented that friends and co-workers often treated him "differently" when they found out that he was a dialysis patient, and he complained, "It's hard—they treat you like you're sick and you're not normal. It's gotten so I don't want to eat my lunch on the job because people look at it and say, 'Why are you eating that?', and so I'd rather just go and eat alone." And finally, a long-term patient noted that a relative had gone so far as to keep her children away from him because of his illness. He said, "She's afraid they may 'catch' kidney disease."

Several patients' family members commented on "stigmatizing" elements of ESRD and dialysis. A mother stated that her son often did not tell friends or co-workers that he was on dialysis and, in fact, did not want them to know. She reported that if they called while he was at a treatment session, she would simply say that he was "out for awhile." Another patient's spouse commented extensively on her husband's initial perception of his condition as "embarrassing": "When he first went on the machine he was very secretive—no one knew where he was except me. Now he's accepted it but it wasn't easy for him to accept. When they told him that's what it was he was very resentful—he diidn't want to talk about it. I'd go to see him and he wouldn't say anything; that went on for about two or three weeks and then finally he started talking and he accepted it."

Hemodialysis caregivers also commented on their perception of the "stigma" attached to ESRD and dialysis. One therapist noted that some patients never told their extended family members (living outside the home) that they were on dialysis. She added, "Some of the patients are very conscious of it and sensitive about it [dialysis]." Another staff therapist reported that patients had actually told her they felt "stigmatized" because of their condition. She stated,

> Some have told me that [they feel stigmatized], but I haven't exactly gotten the reasons. I think it is because they feel they're going to be dealt with differently, and it's frightening to them. It's like a hidden disability. And it's sad. It's like if you had a child that's in a wheelchair, you put it in the closet and leave it there, you know,when people outside the family come around. And that's what they do. They put it in the closet. I did hear that from one gentleman. He said to me, "I feel like people would treat me like I was different, like I was sick," something like that. He didn't want that. He definitely didn't want that.

A social worker focused upon the difficulty that she felt hemodialysis posed for the younger population:

> I think that stigma happens a lot in the younger population. A lot of the young people tend to isolate themselves; and [from talking to them] it seems to come about as a result of reduction in social activities; not being able to go out to eat, not going to a tavern, something like that, and having to restrict their intake of certain foods, as well as changes in physical activity.

Finally, a physician interviewed in the study spoke directly to the issue of stigma:

> Everybody is affected, not only the patient, but the rest of the family. A lot of the patients lose their former friends—their confidence in themselves, you know,

is lost. Some patients even refrain from going out because somehow there is a stigma to the disease although it is not a communicable disease. Just that they are sickly, that they can't keep up with things anymore the way they used to do, I think it is more stigmatizing in males rather than females; because of the role that they used to carry, or the burden that they used to carry.

Overall, the concept of stigma emerged more notably for male study subjects than for females. Its basic referrants to dialysis were the lack of ability to urinate and sexual impotence. Such side effects of the illness condition occasionally seemed to erode a patient's sense of masculinity and spill over into other aspects of social and physical functioning. Role reversal, especially in terms of the "breadwinner" responsibilities, was particularly difficult for the male patient to cope with, and sympathetic responses of concern or pity from family and friends could be and were interpreted as demeaning.

Female dialysis patients seemed to focus on the stigmatizing effects of decreased body image or lack of normalcy because of access scars, inability to pass urine, and decreased libido. Role difficulties were also encountered by females; this was manifested in one patient's report of her child's comment, "Why can't you be like other mothers."

Inability to participate fully in recreational activities with family and friends—particularly in eating, drinking, and traveling—was reported as problematic for both males and females. Some physically disabling effects of the disease, such as weakness and fatigue, were also problems for both sexes.

Inherent in the concept of stigma for dialysis patients, as for any chronically ill person, is a symbolic negative connotation regarding one's wholeness as a person. Patients must not only accept their new and long-term bodily limitations and the impact upon their own life activities and goals but must also cope with living in a society where great value is placed upon health and physical fitness. A hemodialysis patient must continually strive to achieve a degree of physiological "normalcy"—being pressured all the while by values and norms of a "healthconscious" population of peers.

Social Functioning

Social functioning was evaluated quantitatively in the research utilizing the investigator-developed "Inventory of Social Functioning" (see Appendix), and this aspect of the study is discussed also in Chapter 7. Additionally, various elements of social functioning including such factors as role-maintenance and/or role-reorganization were examined qualitatively during the course of focused interviewing.

In a series of articles concerning the adjustment of the chronic hemodialysis patient, Abram suggested that serious conflict occurs between total dependence upon a machine for one's life and the attempt to maintain at least some of the social activities consonant with normal living. He added that the hemodialysis regimen offers a unique opportunity to study man's adaptation to survival by machine and to learn how his life in general is affected by this unprecedented treatment procedure.[21–23]

Much of the extant literature dealing with chronic illness in general, and renal failure in particular, touches upon the patient's necessitated adjustment in regard to social functioning relative to such areas as work,[24–26] recreation,[27,28] family interaction,[29–31] role relationships within the family,[32–35] and attendance at church/club meetings and activities.[36,37] Also discussed is the quality of life for the chronically ill patient, and especially for the hemodialysis patient.[38] Respondents in the present study discussed modifications in social functioning as primarily related to work, recreational, and family-life activities. Sexual functioning was highlighted as an area of concern for most patients.

Work

A number of the study subjects were found to be not working or only working part-time due to their disability or perceived disability associated with ESRD and dialysis. Some hemodialysis patient respondents were, however, gainfully employed in full-time occupations. Statistical analysis of the demographic correlates of work status showed no significant differences. In general, however, those patients who were working either full- or part-time appeared somewhat "healthier and happier" than those who reported themselves unable to work. One "full-time" housewife and mother commented, "I have to keep going because my family expects me to. My children expect their breakfast in the morning and clean clothes and what they's used to; so, I keep up." A male patient who was working full-time in a skilled occupation reported, "I'm more restricted than I want to be. It [dialysis] cut me [activities] back quite a bit, but I don't knock it because it could have been worse. It could always have been worse."

Some patients reported that they could carry a modified work load, and several female respondents confirmed the ability to do light housework but admitted needing help, as one put it, with the "heavy stuff." Most of the patients, especially those longer on dialysis (9–11 years) admitted that very heavy physical work such as heavy housework, yard maintenance, or snow shovelling were now beyond their capabilities. A male patient described his working ability this way: "After a long time at this [dialysis] you have some limitations—you just don't have the strength you used to."

Recreation

Many dialysis patients, especially those who were working full-time, commented that their recreational activities had been decreased due to lack of time as well as certain physical limitations. One patient observed that after a full week of work and three evenings of having hemodialysis treatment, he was often not only too tired to go out on the weekends, but had other things to do, such as laundry and household chores. The patient observed that coming to the hemodialysis unit was, in his opinion, a lot like having an extra job. Some patients also reported that their recreational behavior was now limited to nonphysical kinds of activities such as attending spectator sports functions, watching television, and reading and the like. And certain patients admitted that their recreational activities had decreased notably because of increasing social isolation. These situations were, however, frequently linked to physical disability, with the patient reporting that he or she became fatigued easily or just didn't have the strength and energy to "keep up" with their friends and with former recreational activities.

Several patients—the minority, however—reported the ability to continue to engage in such physical recreational activities as walking, dancing, playing ball, and skiing. Nevertheless, most of these patients also complained that the "machine-time" and trips to and from the hemodialysis unit cut into their recreational schedule, and sometimes hampered their participation in more formally scheduled group activities, such as parties or trips.

Family Life

Ulrich has asserted that, "Role changes frequently accompany the integration of the ESRD medical regimen into the client's life style. As the symptoms become more pronounced, the individual tires more easily and may not be able to participate in work situations or relationships in his usual manner."[39] And, in reporting on a group of male dialysis patients, Cummings noted that, "Three main aspects of the male role in our culture—breadwinner, disciplinarian and decision maker—seem to be particularly vulnerable to encroachments of kidney disease and dialysis."[40] He added that the resultant circumstances "cannot help but demote the patient in his own eyes."[40]

For almost all of the dialysis patient study respondents, social behavior in terms of family-related activities was an aspect of their lives that they attempted to maintain in as "normal" a fashion as possible. Patients generally reported continued involvement with and interest in the activities of their spouses and especially those of their children. One female respondent, a mother of two, noted that she always attended her children's school functions (plays, PTA meetings, picnics) and was involved with their youth groups at her church. Another patient, also a mother of several children, observed that while she did sometimes feel very tired and "washed out," especially immediately after the dialysis treatment, she tried to "keep up with the children's activities" and attended their school and club functions. As the mother of two teenage children expressed it, "I refuse to let that happen [inability to function in her role]. I have a great family and they expect me to go and do the 'motherly' things. I could have sat around and felt sorry for myself, otherwise." And one patient even reported sacrificing her dialysis patient role to that of her role as wife. She described the following incident:

> I don't feel too good today—probably it's because I missed my treatment on Monday. I haven't had one since Friday. I came but they was real late and had problems gettin' people in and my husband is sick and I just couldn't sit around here all day. I had to go home and see about him. His mind is wanderin' and he forgets things— I didn't know if he'd go out or leave the door open—or turn on the gas or something. I don't usually ever miss a treatment. I watched my fluids though and they told me I did okay with my weight, but I didn't feel too good this mornin'.

Male dialysis patients interviewed also reported continued involvement with their children's school and recreational activities. One father noted that he always tried to attend his son's swimming meets; another participated actively in his daughter's college preparatory activities.

Female patients who were mothers generally stated that they attempted to maintain or retain, to the degree possible, their "mother role," with the associated nurturing activities within the family. Patient fathers reported a serious attempt to retain their roles as "head of the household" and sometimes as "disciplinarian." A difficulty for certain husband and father dialysis patients, however, was posed relative to their previous role as "breadwinner." In many cases male respondents admitted that they could no longer financially provide for their families, and some described a situation of role-reversal, with their spouse now handling the major financial, physical, and social-interactional responsibilities of a household head. One patient stated, "My wife has taken over the household—money, everything—there's not too much I can do." Another male patient and father of three asserted that, while he could not financially and physically provide for all of the needs of his spouse and children, he felt that it was very important that he maintain his role as "father" and "head of the household." He noted, "That's why I try to survive as long as I can, because I don't know how they would live without me."

Sexual Functioning

Any research dealing with the social and social-psychological adjustment of the chronic hemodialysis patient must consider the topic of sexual functioning. It has frequently been reported in the literature that end-stage renal disease and its treatment modality of hemodialysis may result in partial or complete impotence in males, and cessation of the menses and lack of sexual desire on the part of the female.

In discussing the problem of stress for the dialysis patient, Crammond stated that while there are various ways of coping, "the sexual outlet is usually blocked by the loss of libido which accompanies the condition."[41]

Levy, who conducted a national survey on sexual behavior among hemodialysis and transplant patients, reported a 59 percent response of total or
partial impotence for male dialysis patients and a marked decrease in the
experiencing of orgasm for females.[42] The responses were controlled for
degree of sexual functioning before the development of uremia and insti-
tution of the hemodialysis procedure. In summarizing, the author pointed
out that, "hemodialysis patients of both sexes and male transplant recipients
experienced great deterioration in sexual function at the time of answering
the questionnaire in comparison to that prior to their development of uremia.[43] In working with hemodialysis and transplant patients at a military
medical facility, Collins discovered that, "inadequate sexual responses were
frequently reported by patients and their spouses," and pointed out that a
number of male patients were impotent.[44] These reports of sexual dysfunction among end-stage renal disease dialysis patients are also supported by
the findings of Milne et al, who noted further that despite the prevalence
of sexual problems among hemodialysis patients, "the number of those requesting help was notably lower than in other populations."[45]

Because the investigator in the present study anticipated that some
study subjects might not be at ease in discussing this behavioral area, sexual
functioning was treated near the end of the initial interview schedule. Very
few respondents, however, refused to discuss this subject area; and a sub-
stantial proportion were, in fact, quite verbal, asserting that they felt impotence (males) or loss of libido (females) was one of the major problems
associated with their illness.

Of the total patient population interviewed at Time 1 (N-126), only 4
patients (3.17 percent) had no comment when questioned on this matter;
22 (17.5 percent) said there had been no notable change in their sexual
functioning; 49 (38.9 percent) reported a decrease in performance; and 51
(40.5 percent) stated that they had experienced total impotence or loss of
libido.

In analyzing the above findings according to sex, it was found that although the overall distribution of responses were approximately similar,
more males, 36 (48.0 percent), reported total loss of sexual functioning
than did females, 15 (29.4 percent). A higher percentage of white respondents, 22 (50.0 percent) than black, 29 (35.0 percent), reported total loss
of sexual functioning. Blacks, however, showed a higher incidence in the
category "decrease in performance or desire," with 36 (43.9 percent) responding affirmatively, compared to 13 (29.5 percent) of the whites. In
regard to marital status, data revealed a notably higher percentage of "total
loss," 41.3 percent, reported by married patients than by patients reporting
themselves "never maried,"—23.5 percent of the study respondents.[46]

Finklestein, et al. suggest that sexual functioning must be examined in
terms of how dialysis affects the marital relationship, and report their finding

that there is a problem "if one looks at the desired (as opposed to the actual) frequency of intercourse."[47] And Alexander and MacElveen note that in the marital situation, "the sexual relationship can be vastly altered depending on how the patient feels."[48]

When dialysis patient respondents in the present study were initially asked about sexual functioning, informal comments expressed the degree of their personal concerns:

A. (Male patient, married, age 56.) "There is no sexual potency. I've been wondering if you would ask me this question. It's [impotence] one of the things I've felt bad about—my wife is still a young woman. She says it doesn't matter but I know better."

B. (Male patient, married, age 36.) "I'm almost totally impotent since I've been on the machine. This is very difficult for my wife . . . I can't keep her in a prison."

C. (Male patient, married, age 58.) "There is no potency—it's a pain—a woman can say it's not a problem, but a man can't say this because he's the one who has to perform. I wouldn't even think about touching a woman."

D. (Female patient, married, age 30.) "My sexual appetite has been nil for some years . . . I just don't have any desire . . . my hair style and clothes have gone downhill too . . . I don't seem to care much about my appearance.

E. (Female patient, separated, age 39.) "I have no sex desire at all—haven't had any for the past two years. I was afraid when I heard that sex was a problem with this condition . . . that this might make me a freak."

F. (Male patient, married, age 52.) "Why should I comment [on sexual functioning]? You know darn well there isn't any and anyone who says there is is a liar. I'm totally impotent.[49]

In looking at changes overtime on the item in the study's quantitative interview schedule measuring sexual functioning, the data can be summarized for the 63 respondents interviewed at Time 2 as follows: 27 patients (43 percent) reported a decrease in sexual functioning, the majority changing from the response category "decrease in performance or desire" to "total loss of potency or libido"; 6 patients (9.5 percent) reported an increase in sexual behavior, progressing either from "total loss" to "some decrease in functioning" or to the response category of "no problems"; and 29 (46 percent) dialysis patients' responses remained unchanged over the 3-year interval. This latter group consisted of 4 patients (6 percent) reporting "no problems" with sexual functioning; 9 (14 percent) admitting to a decrease in performance or desire; and 16 (25 percent) reporting "total loss" of sexual potency or libido. One patient who had declined to comment during the initial interview (T1), reported total loss of sexual functioning at Time 2.

At the time of the third interview (T3), sexual functioning appeared to have remained relatively stable in the previously established categories for most of the 33 patients interviewed. Several of the patients did, however, discuss the difficulties related to the sexual problems that they experienced over time. One male patient commented on the stress that his impotence caused for his wife, and another noted that lack of an active sex life made him feel "not normal." A few patients were quite positive about their sexual functioning. One respondent discussed "the psychological problems" related to sexual behavior that she had perceived some dialysis patients to have. This patient related that another dialysis patient had told her early in her illness that she would not be able to function sexually once hemodialysis was initiated. The respondent recalled the incident this way:

> She messed me up—messed up my mind. She said, "I'll tell you one thing, you can't have any more sex once you're on the machine—that's just dead." And I was so upset I didn't say anything to anyone, but there I was, a young married woman with a young husband. This lady told me she had a friend on dialysis who told her that she would pretend that she could do something but really she was dead. I wanted to ask a nurse about that, but I couldn't. So, my husband and I, we really had to get ourselves together and get straight. But I found out it's not true about sex—it's all in your head—the problem is in your head.

This patient reported enjoying an active and satisfying sexual relationship at the time of interview.

RELATIONSHIPS WITH SIGNIFICANT OTHERS

Family and Friendship Relationships

Family

Sorenson reported that one source of tension in dialysis patient–family interaction lay with the spouses who assume the primary responsibility of caring for the patient.[50] Similarly, Bailey reported that when a wife learned to monitor the treatment procedure for her spouse, role reversal was initiated as the former breadwinner became dependent upon his wife for the continuance of his life.[51] Such stressful situations can notably alter the usual pattern of a husband–wife relationship. Findings reflective of the change in family relationships were varied among the present study groups.

In terms of the quality of interaction with family and close friends, a number of patients reported either the continuation of previously stable relationships or noted the initiation of more positive interaction related to the care and support provided relative to their illness condition; the majority, however, suggested in their responses that some deterioration had occured over time. Quantitative data elicited in structured interview sessions also revealed that "quality of patient interaction decreased moderately between initial (T1) and follow-up measurements."[52]

The expectations of family members provided much support and sometimes comfort in terms of the patient's functioning. One female patient who was living in an extended family situation observed, "I wash and cook and go back and forth to the store. They [the family] don't expect me to sit down just because I'm sick. But they're here to do for me when I can't."

Family support was described in various terms by hemodialysis patients—sometimes in terms of direct comfort and caring, sometimes in terms of actual services performed, and sometimes in terms of family expectations and needs as they were interpreted as providing the dialysis patient a reason for living. A male respondent focused upon the aspect of caring in his comment: "My wife is really a good woman. She's stuck by me all through the years. Another woman would have let me go a long time ago, but not her. She worries about me all the time, especially if I'm sick. We stay . . . always [close together]." Another patient commented on the support of being married and having a child: "I'd be long gone if I lived alone. First of all, my wife gives me a lot of love. My daughter and I talk a lot too and she'll show me her school work when she comes home. It keeps me living." A wife and mother patient stated, "It [your family] makes you go on and do what you have to do. Without them, what would you have to live for." And another patient mother commented on the support and caring received from her children:

My daughters have been wonderful—they have been right there with me all the way—if I'm not feelin' well, one of them stays home and comes to the dialysis unit with me. I never miss, even in the snow. I have one daughter who was only 2 years old when I started; now she's 10. I have explained it all to her [dialysis]. I have even brought her out here to see how the machine works. All my daughters have been here except my oldest. My daughters really watch after my diet, especially my oldest. If I eat something wrong she will just have a fit and say, "What are you doin' with that?" Then they threaten to call the doctor. And don't let me sit like this holdin' my head for a minute. I'd be sittin' like this and they right away say, "Oh, Momma, what's wrong?" Sometime if I don't feel like gettin' dressed, they dress me—I say, "Oh, don't you all pamper me so much."

Other patients commented on such things as family members taking over heavier physical activities like driving or cleaning the house. One female patient reported, "If my husband sees something to be done [at home], he just goes ahead and does it. It's hard on him but he does it anyway." Another added, "I credit my husband with my adjusting so well. When my kidneys first failed, I was out of it and he took over everything."

Several patients commented, as noted earlier, that they were helped by the fact that family members did expect them to carry out their role responsibilities and didn't "baby" them or appear oversolicitous. One male patient stated, "They [family] know I am sick, but I never accept sympathy. I maintain my father role. She [wife] is concerned, but she doesn't let it show." And many patients discussed their positive perception of family support in terms of compliance with the treatment regimen. The following story related by a male dialysis patient exemplifies the point: "The other day we saw some watermelon on the way to church and my wife said, 'Keep on driving, you can't have none of that.' I guess I'd be in trouble all the time if it weren't for my wife. She'll take things away from me in a minute. She don't cook nothin' I ain't supposed to have." The family members' support of patient compliance was not always so overt, however. One patient mother shared a delightful perception about her two young daughters and their insight into her dietary needs and restrictions. She reported, "They don't say much about what I eat, but they know. Sometimes I get a craving for something like banana pudding. I make a bowl and put it in the refrigerator but after I've had one dish, it all suddenly disappears. Then they know I can't have another [dish of pudding]. It's always a mystery where it went."

While, in the main, the dialysis patients interviewed responded positively in regard to their perceptions of support by family members, occasionally such attention was reported to be problematic or even totally absent. One patient who was, at the time of the study, living with friends, commented, "They [friends] are better than family—your family can wear you out with their caring." Another patient observed that his family, with whom he lived, really didn't understand his condition and were not supportive. He suggested that family members had their own concerns which did not include him and noted, "Where I'm at, there's a young daughter and a baby and I'm different—it's not that I want to be, but it's just like that. If they want to party or somethin', I get in the way."

Friends

Some dialysis patients reported receiving much comfort from friendship relationships while others appeared to receive little if any such support. A few patients stated that their friends, acquaintances, and/or co-workers did not even know they were dialysis patients and they (the patients) liked it that way. One respondent expressed it this way: "Ninety-five percent of my friends don't even know about this [dialysis]—that I come here [to the unit]. They don't really need to know about all this—you have a private life. Most of them probably wouldn't even believe it even if I told them about all this—they wouldn't think I look sick."

In certain cases, however, a dialysis patient's primary "significant other" was categorized under the label "friend." Sometimes the phrase "person I live with" was used by the patient to reference the identified other, generally indicating a relationship with no formal legal or kinship-tie implications. One female patient described her situation this way:

I have a chronic illness and I have to come in on this machine three times every week or I'll die. My family don't really understand this, or at least they pretend not to. But M., she's the woman I live with, really understands me. M. is real good to me. She made me feel very good the other week. She said, "A., I love doing what

I do for you—I don't want anything for taking care of you." She is a good help. If I say I want a coke, she says, "No"—if I get some cookies, she says, "Give me them and I'll get rid of them."

The patient added, "I really love her a lot."

A male patient also reported the importance of the support and caring of a friend whom he described as "T.—my buddy," and added, "He stays with me." The patient noted that T. took him to the unit for treatments and picked him up, and sometimes stayed around if he was not feeling well. The respondent reported, "I couldn't make it without him."

In hemodialysis patients' relationships with family and friends, the following pattern of interpretative attitudes and behavior predominate:

1. Family/friends don't know about (the patient's) dialysis.
2. Family/friends know (about dialysis) but don't treat the patient as ill.
3. Family/friends know (about dialysis) and treat patient as ill.
4. Family/friends know (about dialysis) and abandon patient.

RELATIONSHIPS WITH CAREGIVERS

It has been noted in the literature that the behavior of chronic dialysis patients is strongly influenced by the attitudes of those in the caregiver role.[53] Roper, Raulston, and Cramer have suggested that staff–patient rapport is a primary factor in the long-term adaptation of the maintenance dialysis patient.[54]

Hemodialysis patients' relationships with their professional and paraprofessional caregivers in this study were found to be of both an instrumental and an affective nature. Some patients reported these relationships to be primarily instrumental, dealing with caregivers only in relation to the dialysis treatment regimen. The majority of patients, however, stated that they had developed close and caring relationships with at least one or two of the dialysis unit team members over the course of their hemodialysis history. It was the existence of these relationships, in fact, that caused dialysis patients to worry and sometimes agonize over the problems of "staff turnover" as discussed in Chapter 5. One respondent observed that usually the patient group got so close to the staff that it caused a great deal of anxiety when a caregiver left, and added that a lot of discussion was carried on relative to the staff member's replacement. Another patient admitted, "I got very close to one therapist. You get to know them and they know you and you really feel that you can trust them for the machine. Then, too, you can talk to them about your problems, but you can't do that with everyone." Many patients expressed having mixed experiences with caregivers, as the following patient spouse's statements exemplify:

Everyone [on staff] has been so good and we really have gotten close to them over the years. Of course, there are always a few that are difficult. And we're always glad to get away from the unit because of those few. They have really made life miserable at times. I know A. can be obnoxious sometimes, but for the four hours he's there, it seems as if they could put up with him. We've met a lot of really nice people but we've also met some "dillies!"

In regard to their patient–physician relationships, most patients had positive comments. The most notable complaint was that related to availability or accessibility of the physician. As one patient commented, "One complaint I have about this place is that they [the physicians] only come on Friday. And the doctors alternate. I think a personal relationship with the physician is so important." Several patients commented that they felt that the personnel should spend more time with patients, and one observed that staff members could "be more compassionate of the patient's feelings." One patient, however, supported the staff and observed, "They would spend more time talking to us if they could, but they're overworked and that's not good. If they're tired and all upset and working too hard, they can't take care of us like they should."

RELATIONSHIPS WITH OTHER DIALYSIS PATIENTS

The concept of dialysis patient-to-patient relationships has not been examined extensively in the literature, yet both caregivers and patients themselves note that many of these relationships do exist. It is often reported that such relationships may provide support for the newly adapting dialysis patient. Gelman, a member of a dialysis center's department of social services, admits "as helpful as we can be in aiding new patients in their adjustment to dialysis, there's nothing quite like having them get some first-hand advice from someone who's not only 'been there' but is 'still there'."[55] Mulkerne and Tucker, in discussing the concept of peer facilitation with hemodialysis outpatients, report, "The peer facilitation training has not only provided a useful tool for assisting patients with adjustments to dialysis problems, but has also enhanced aspects of the peer facilitators' lives."[56]

Number and type of patient–patient relationships reported by the study respondents varied. The waiting room was cited by many patients as the place where patients interacted and got to know each other. One male patient stated that he got to know many other patients because of time spent in the waiting room. He added, "Quite a few of us talk together—about dialysis and about our problems. We compare notes and it's nice to have somebody who understands and is going through what you are." Another patient described the waiting room as "the patients' turf" and observed, "We can just be ourselves because it's all dialysis patients—sometimes the families are there, too—and we talk about dialysis or the staff or whatever."

Both the patients themselves and caregivers frequently described how patients "watch each other" in the unit during the treatment procedure. One patient observed, "We watch out for each other on the machine. We take care of each other." A caregiver commented, 'These patients really do watch each other's machines. They'll tell us [staff] if a trap is low or if somebody doesn't look too good. They notice everything." A nurse added, "They definitely watch each other. In fact, we have chronic patients who ask where other patients are, even if the patient is dialyzed in the other room. We have two rooms, and if they are usually dialyzed in the same room together and then they're moved to another room, they'll always ask if that patient is okay, if that patient is here today. They watch each other like hawks." The patients' "watching of each other" was also described as being carried beyond the treatment procedure. One dialysis unit charge nurse observed, "These patients watch each other all the time. Like if somebody goes on CAPD. They watch to see how patients from other shifts are doing on new procedures and on transplant. Then they make their own decision."

It is suggested that ESRD and hemodialysis might be considered an "equalizer" among patients of various socioeconomic and/or cultural backgrounds; this is validated by an example provided by a hemodialysis unit head nurse. In discussing a dialysis unit friendship that developed between two patients of extremely different cultural and economic backgrounds, she stated, "Their life-style was so different. And the two of them would just talk about their diet and the protein business and it was really strange because they always found a whole lot to talk about and were really friendly. . . . One time I was talking to her and she was going on about her Persian rugs and I was talking to him one day and he was telling me what it was like to sell drugs on the street."

Despite such varied backgrounds, many patients reported some interaction with other patients outside the unit. One patient said, "We call each other and sometimes we get together and have parties. You know they got me chocolates and came to see me when I was in the hospital." Another patient described the development of a relationship this way: "This one guy he was a big government official and he was frightened and I talked with him. He calls me a lot now."

The concept of helping new patients was sometimes verbalized by long-term dialysis patients in discussing their patient–patient relationships. One patient explained it as follows:

I try to talk to the new patients. A lot of patients come on this machine now. If I sit next to somebody new, it gives me somethin' to do, to pass the time. I talk to people more now than I used to. I try to tell them the kinds of things that I went through. The length of time that I been on the machine—it's not so bad after you get used to it. You keep talkin' to them and it kind of helps. There's this new girl on the machine in the unit. She asks me questions and I tell her things. I tell them it's not as bad as it seems. When they see me, how I'm doin', they say, "I guess he knows what he's talkin' about." After so long—it's been about 8 years now. One thing I tell people, if you haven't ever been on the machine, you don't really know. You have to have done it to really understand. I like to talk to other people now.

Another patient commented on trying to help other dialysis patients, stating, "I talk a lot to the other patients at the center. I can tell them things that help me when I'm on the machine and they probably tell me some things that help them. I told this man who was upchucking on the machine to try to eat something before the treatment and it helped him to toughen up."

One other way for the development of patient–patient relationships to occur is through some type of formal patient or patient/family support group such as NAPHT (National Association of Patients on Hemodialysis and Transplantation) or a local type of club or association. One study respondent spoke expansively of her activities in a local kidney patient group and seemed to find much positive benefit and suport from her efforts to work with other dialysis patients in fostering the group's development and growth.

DIALYSIS PATIENT FRIENDS' DEATHS: "PULLING BACK"

A fairly common theme that emerged in discussions relating to inter-action with other dialysis patients was that of patient deaths. It has been suggested that, "probably one of the greatest underlying stresses faced by dialysis patients is their fear of death. Self-preservation is the most basic human drive. The dialysis machine is a constant reminder that the client belongs to two worlds: the world of the living and the world of the dying."[57]

A number of dialysis patient study respondents admitted that they worried about their "dialysis patient friends," especially if the friend's physical condition began to deteriorate in any way. Several patients admitted that there was a strong "identification" because of the relatively similar physical condition of the patient friends, and thus death of a dialysis patient friend was personally threatening as well as painful. The surviving patient experienced not only the loss of a comrade with whom he or she had been able to share, but also a loss of security relative to his or her own future. Because of the seriousness and personally threatening nature of these types of death-losses, the surviving dialysis patient sometimes verbalized an attitude or behavior of pulling back from any kind of personal involvement or friendship with other dialysis patients. One male patient respondent reported that he had several times gotten very close to dialysis patients who had died and admitted that he had ultimately decided not to get involved again because the losses were too difficult. He described his experience and present feelings this way: "I'm just not going to get close to these patients again. I really have with several. I've sat next to several patients that have passed, and the last one was very bad because we had gotten to know each other real well—for a couple of years. It really hurt when it [the death] happened and it's scary. I got despondent. So I just decided, no more." Another patient reported that he had gotten extremely close to a patient who had many physical problems as well as ESRD and that the respondent used to help and encourage him. The patient respondent stated, "I know he missed me when I was in the hospital. We got real close. He would call me every night. Then he died. After that I try not to get too close to anybody." One female patient commented on how difficult she felt it was when patients in her own unit died, because most of them knew each other fairly well. She observed, "About 10 patients died in just one year. It was so bad the doctors sent someone around to ask us how we felt about it. What can you say? You feel terrible!" The respondent then added, as an afterthought, "You know, they never said it [the cause of death] was 'the kidney'; they always said it was the heart or a stroke or something like that."

A family member commented on the way patient death affected her family-member dialysis patient and reported, "C. gets really emotional if something happens to one of the kidney patients—especially if someone dies."

Many caregivers discussed the effect of dialysis patient deaths on their unit's patient group. One therapist noted that patient deaths so upset others in the dialysis unit that when possible the staff tried to call patients at home to inform them about a death personally and "prepare them," rather than have them find out when they came for treatment that someone was gone. She said, "They're a very close-knit bunch. They just want to know what happened. They want to reassure themselves that they don't have the same problems that 'did in' the other person. They want to know if the death was related to their [the dead patient's] kidney failure, their dialysis; if they had some other disease that they themselves don't have." The latter portion of the above comment, relating to a surviving patient's concern about the cause of another patient's death, describes a commonly heard theme. As a long-term dialysis therapist put it, "When patients die, other patients want to be reassured that it wasn't the kidney disease that killed them. They like to hear that it was their heart or an accident and anything but the kidney that killed them." Patients were very upset if another patient died and they were not told the reason. A patient respondent commented, "We need to know these things." A caregiver observed that after a patient death, the surviving patients reflected the negative impact on them. She described it this way: "You can almost feel the feeling of fear in the other patients. It's almost like a withdrawal with most of them. They become very quiet. And you know it may take a week or so before they come out of it."

Finally, a hemodialysis unit social worker seemed to sum up the patients' reaction well in her example of one patient with whom she had worked. The patient was quoted directly as saying that after experiencing a dialysis patient friend's recent brush with death, "I will never get close to another patient again. I can't handle seeing it happen to that girl again or anybody else. I've got to get away from here. I've kidded myself that I've really made good friends with these patients. But I can't. I've got to go on with my own life. I can't watch this day in and day out."

Dialysis patient respondents in the present study, though fairly open in discussing the deaths of other patients, rarely mentioned or even alluded to the possibility of their own deaths. It appeared that a larger measure of denial of one's own physical fragility had been instituted as a means of coping with the day-to-day stresses of the illness condition and its treatment. For ethical reasons, it was determined that the concept of personal death should not be pursued in open-ended interviews.

RELIGION AND SPIRITUALITY*

In the research the support of religious and/or ethical beliefs in coping with renal failure and dialysis was examined in some depth. Data are representative of the patient respondents' perception of the import of these beliefs and affiliations in regard to adaptation to their illness condition and treatment regimen.

There is a paucity of literature dealing specifically with the association between religious orientation or beliefs and adjustment to long-term illness; however, certain sociologists of medicine have addressed related issues. Friedson suggested that historically the rise of Christianity "changed the definition of illness from a generally naturalistic one to a religious and a supernatural one."[58] It has frequently been observed that persons faced with serious illness turn to religion as a source of comfort and peace. The ill, as discussed by Susser and Watson, nevertheless showed certain variability in their religious responses.[59] David Mechanic, in a study of illness behavior among Catholic, Protestant, and Jewish college students, reported a greater incidence of illness among Jews than among either Catholics or Protestants; the study, which also controlled for subjects' income, suggested that differences in illness behavior were most significant, however, for the higher income group.[60]

*Major portions of the following discussion are reprinted from O'Brien ME: Religious faith and adjustment to long-term hemodialysis. J Relig Health 21:68–80, 1982, with permission of the publisher.

Relative to the chronic hemodialysis patient, upon whom much research has been focused in regard to the physical, psychological, and sociological correlates of the treatment procedure, little attention has been paid to the spiritual or religious dimension of attitudes and behavior. For the end-stage renal patients, who live totally dependent upon a machine for their continued existence, the quality of life becomes questionable and often poses ethical or spiritual dilemmas for both the patients and their families. Certain patients become extremely depressed, sometimes questioning the reason for their plight, as well as the justification for their continuing to live.[61-63] Patients may also become notably more alienated as the length of time on dialysis increases,[64] and many feel very strongly the need for both emotional and spiritual support systems.

In a study of 21 hemodialysis patients carried out over a two-year period, Foster, Cohn, and McKegney reported that all Catholic study patients (N = 9) survived and the only Jewish patient died.[65] The authors appear to consider religion as a significant factor but did not discuss its implications theoretically. Barry, in a study of 7 dialysis patients, related numerous positive responses reported by individuals in regard to the influence of religion; data indicated that these patients might not have adapted to their illness and its treatment regimen without the strength provided by their religious faith.[66]

In the present study the religious affiliation profile reported at T1 (N = 126) included 25 Catholic, 87 Protestants, and 8 Jewish patients, and 6 subjects with no particular religious group identification. At T2 (N = 63) 50 of the study subjects were Protestant; 10, Catholic; 3, Jewish; and none reported having no religious group affiliation at this time. The variable of religion was defined by items measuring religious affiliation, participation in formal religious services, and both quantitative and qualitative questions reflective of the patient's perception of the import of religious faith in adjusting to end-stage renal failure and the hemodialysis treatment regimen.

Patients' Perception of the Import of Religious Faith

In order to determine whether subjects themselves perceived religion as being associated with their acceptance of chronic renal failure and adjustment to its treatment regimen, dialysis patients were asked to rank order the importance of religious or ethical beliefs in their lives. First, members of the study population were asked to rate the import of religious faith for the illness adjustment process according to the categories "always," "usually," "sometimes," or "never." They were then asked to comment on the preceding answer in as much detail as they desired. Most of the dialysis patients were amenable to discussing at length their perception of the meaning of religious faith, and many held the opinion that religion was of notable import for their adjustment to renal failure and the dialysis regimen.

Overall, 33 respondents reported that religious beliefs were never relevant in relation to acceptance of their condition; 27 asserted that religion was sometimes important; 31 responded, usually, and 35 stated that their religious faith was always an important element associated with adjustment to their illness condition. Thus, 93 of the dialysis study group at T1 held the opinion that religious or ethical beliefs were to some degree associated with acceptance of their disease and its total treatment regimen.

When comparing religious affiliation with perception of the import of religious faith, it was found that 5 members of the total Jewish group viewed religion as never related to adjustment, while only 3 Catholics and 20 Protestants had a similar attitude. On the other end of the scale, 9 Catholics considered religion as always relevant; 25 Protestants reported the same; and one Jewish respondent held this opinion.

Qualitative Panel Data Reflective of Patients' Perceptions of the Import of Religious Faith

In content-analyzing qualitative data on individual cases for change over time, it was found that 18 of the patient panel group (N-63) at T2 suggested in their responses that their perception of the import of religious faith in adjusting to their illness had either changed from a negative to a positive attitude or had notably increased in the degree of import perceived during the previous three years. Of these 18 respondents, 10 had changed their response from one indicating that religion was of no influence in their adjustment to their illness condition to a strongly positive perception of the import of their religious faith in this area. One patient, who had asserted initially (T1) that he did not feel religious faith had any influence in adjustment to his condition, stated at T2 that he had "found the Lord" through his illness. He added that, during three cardiac arrests and numerous surgeries, he had prayed, and he noted, "Each time I knew everything would be all right because I asked God to carry me through—I know that He's got His arms around me." Another patient who had previously stated (T1) "the church didn't help" and "religion does not influence me," at T2 responded, "without that [religion], I would have no faith in dialysis or the people working with me—I'm just beginning to accept." A third patient who had simply replied "none" to the question of the import of religious faith at T1, stated at T2, "Oh, yes. A lot of people couldn't have gone through what I went through without faith in God." Another patient, also having reported a negative attitude toward religion at T1, stated at T2, "I have grown a lot closer to God [through this experience with chronic renal failure]."

Only one sample respondent changed from a positive response regarding the import of religious faith at T1 to a negative perception at T2. This subject stated his rationale for the T2 answer as being related to the fact that he no longer "went to church." At the time of initial interview the patient had placed much importance upon going to church and stated, "It makes me feel relaxed."

In examining sociodemographic factors in regard to changes over time in religious faith, it was found that data were variable for those patients (N-18) whose perception of the importance of religion in adjustment to renal failure notably increased over time: 8 were men, 10 women; 14 were Protestants, and 4 were Catholic. Ages ranged from 38 to 71, with the modal age being approximately 45. Educational levels ranged from fifth grade to bachelor's degree; patients had begun dialysis from four to five years previous to interview at T2, and seven of the group lived alone, the others variably distributed in households with other adults or adults and children. The one patient subject whose perception of the importance of religious faith changed from positive to negative at T2 was Catholic, 52 years old, and had been on dialysis for approximately five and one-half years. This respondent did report that he had some major physical problems during the past three years and would like to have a transplant but was very "afraid of so many operations." He also appeared quite alienated and depressed at the time of reinterview.

Several of the hemodialysis study subjects mentioned that having church members and/or their minister coming to visit and praying for them "helped a lot" in facing their condition, and one respondent stated that religious faith "helped me accept illness" and made a difference in "how I'm living." Relative to alienation, another patient noted that faith in God "makes you feel like you're not friendless and not lonely—you forget all your fears." Another asserted that "it is very important for people who are suffering like we are to have faith." In terms of treatment regimen compliance, a dialysis patient asserted, "I pray to God all the time to help me stay on my treatment and to do what I have to do." Several subjects responded in a similar vein: "Without my religious faith, I couldn't make it." "It [religious belief] really helps you to go on." "If it hadn't been for my religion I wouldn't even be here now." And "Without it [faith] I don't know what I'd do."

Quantitative Panel Data Reflective of Patients' Religious Behavior and Perceptions of the Importance of Religious Faith

In analyzing changes over time for the dialysis patient panel group relative to a quantitative item measuring perceived importance of religious faith in adjusting to their illness condition, it was found that 25 of the respondents became more positive in their scale item response. Twenty-eight of the subjects responded exactly as they had at T1 (three years earlier), and only 10 reported a perceived decrease in the importance of religion in their lives. Of the 18 patients whose qualitative responses indicated a notable increase in perception of the importance of religious beliefs, the majority had correspondingly notable increases on the quantitatively scaled religion item. For the one respondent whose qualitative response signified a decrease in importance of religion, the quantitative response decreased similarly.

Religious affiliation remained the same over time for all subjects but one, this patient indicating a change of denomination within the Protestant group. The respondent's rationale for this transfer was a feeling of need for "more warmth and support from church members."

In regard to frequency of church attendance, 24 of the panel group maintained a similar pattern of attendance at T2, as at the time of initial interview. Twenty-one persons reported attending church services more frequently, and 18 decreased to some degree their church attendance. Ten of those subjects in the "decreased" category reported that they had been prevented from going to church because of physical limitations imposed by their condition (tiredness, etc.). Church attendance was variable for those 18 patients whose qualitative response indicated an increase in the impor- tance of religious faith, with 7 remaining stable, 8 reporting an increase in attendance at church services, and 3 reporting a decrease.

For many hemodialysis patients the condition of end-stage renal failure and the complex treatment modality of hemodialysis might have been less well coped with, by their own reports, without the support of religious or ethical beliefs. Also of influence was the supportive environment provided by members of the patients' formal religious groups. The wife of a dialysis patient expresssed it this way: "I'm sure people who are stricken with kidney disease have faith. If they don't have it before, I'm sure they'd have to get some afterwards—just to help them survive from day to day."

It is interesting to note that representatives of organized religion were largely absent from the dialysis unit itself. While most health care systems dealing with serious illness conditions provide some context for the input of religious functionaries, only one unit in the present study included the regular pastoral counseling services of a single Protestant chaplain. An ex- planation for this might have to do with the fact that the primary focus of the chronic hemodialysis unit is upon life (as opposed to death) and survival. Some healthy denial of the seriousness of the patients' conditions must be instituted in order to allow the day-to-day business of the treatment pro- cedure to be conducted with some semblance of "normalcy."

MODIFICATION OF LIFE GOALS

Some literature on vocational adjustment in the rehabilitation process among dialysis patients indirectly focuses upon the modification of life or career goals for the patient. Calsyn et al., in a study of 107 hemodialysis patients, discovered a trend that indicated "that patients with additional serious medical problems are more likely not to return to work."[67] Goldberg, in reporting a study of vocational plans of 27 male dialysis patients, noted that the "group as a whole significantly lowered their acceptance of responsibility for vocational plans during dialysis."[68]

While some study dialysis patients were able to maintain professional career activities and, as previously noted, family roles and role responsibilities, many patients expressed the frustration of having to modify or surrender previously envisioned life goals. Often the goals that patients reported they now either could no longer plan to achieve or could possibly achieve only to a partial degree, related to the areas of occupation and education. One patient commented, "I always thought I'd like to go into higher education, but I can't do that now." Another stated, "There's a lot of things I haven't been able to do because of kidney failure. I wanted to be a hairdresser, but because of this I couldn't do it. I couldn't stand that long now." Several male patients reported that they were forced to give up their jobs, especially if any manual labor was involved, because they were not physically able to handle the work and this was to some very frustrating. Those patients who had achieved certain initial life-goals in regard to education, raising a family, or entering a career (particularly the more intellectually or technically oriented careers) were more successful in continuing their activities and were more satisfied with the life modifications imposed by ESRD and the hemodialysis treatment regimen than those patients whose attainment of their goals remained still in an uncertain future.

QUALITY OF LIFE

"Quality of life" is difficult to define for any group of people, much less the chronic hemodialysis population who must live their lives dependent upon a machine for continued survival. In discussing a study that examined life satisfaction among renal dialysis patients, Jackle noted that "life satisfaction— a person's general appraisal of his life—is considered by professionals and laymen alike to be a major component of any comprehensive view of the quality of life."[69] Thus, life satisfaction, as an indicator of quality of life, was evaluated in an attempt to establish the patients' feelings about their lives. When compared with a normative population, a group of 30 dialysis patients placed themselves on the average at about the midway point on a life-satisfaction scale, i.e., "between the best and the worst possible life; the average member of the normative group saw himself about a step and a half higher on the ladder [scale]."[70]

Quality of life in end-stage renal failure tends to surface in discussion whenever decisions must be made to initiate, continue or terminate treatment; when gross noncompliance indicates a possible death wish on the part of a dialysis patient; or when frustrated dialysis caregivers question the treatment wishes of a patient's family or kin. It has been pointed out that any illness has a decided impact upon quality of life, and especially noted is the "impact of interpersonal relationships . . . [and] interference with individual freedom."[71] Gutch and Stoner list three implications of quality of life for the dialysis patient, described as relating to (1) withdrawal from treatment [active or passive suicide], (2) intense stress upon the family group, and (3) stress on dialysis unit caregivers "for whom failure by a patient is somehow equated with professinal failure."[72] After studying medical teams in a number of dialysis units, Kaplan De-Nour and Czaczkes reported their impression that there seems to be a "unit opinion about dialysis and patients undergoing this treatment.[73] They explained that staff members in different dialysis units appeared either to view life on dialysis as a terrible thing, with patients being miserable, or to see hemodialysis as a marvelous procedure, resulting in happy and well-adjusted patients.[73]

It is interesting to note that interviews in the present study revealed that both caregivers and patients and their significant others viewed quality of life for the chronic dialysis patient as linked to patient "acceptance" of their condition and the treatment procedure, rather than being directly related to ESRD and the psychosocial limitations of the hemodialysis regimen.

Overall, dialysis patient respondents suggested that the quality of their lives was good or at least, not "too bad." Some, of course, did perceive a negative impact upon the quality of their lives related to ESRD and dialysis, but this was often associated with severe and/or recent physical deterioration. The patients themselves expressed the importance of "accepting" their condition. One young male patient, who was working full-time in a professional capacity, described it this way:

> You have to pull yourself up and do for yourself. You can't keep waiting for everybody else. Sometimes when I get up in the morning I feel bad. But I just get up and go to my business and make myself keep busy. You have to accept the fact that your kidneys are gone but you can still do things. A lot of it's in your own head—how you feel about it—how you accept it.

Another male patient commented, "this life is what you make it. You can't fight it and be angry and say 'why me' all the time. That gets you nowhere. You just have to accept and do the best you can." A female patient asserted, "I'm happy and grateful for the life I have. I could sit around and mope all day, but where would that get me? I have my husband and my children and they are everything to me. Sure this is hard sometimes, but I'm grateful for the life I have. It could be so much worse." Finally, a male patient who often worked "more than full-time" each week commented on quality of life and patient attitudes and activities in this way:

> I think a lot of dialysis patients' problems are psychological. Sure, after a treatment you can be tired and feel like sleeping, but you have to make yourself get up and

go. I go out to a party or something like that and then I forget I was tired. Some people just let themselves be sick and life isn't worth living like that.

Dialysis caregivers' perceptions and experiences varied on the issue of patient quality of life. Some were positive, as the following statement illustrates:

Quality of life is whatever that patient wants to make it. I have a patient; he had a beautiful home, daughters, a wife in school, the whole bit. Now he is unable to feed himself. Do you know what, he has positive quality of life. They worked really hard to get him a wheelchair with a motor on it so he can get around and he is doing church work. He isn't even sick. You know, that is quality of life. He is very much into it and his family is, too. It is beautiful.

One nurse commented on the relative nature of quality of life for individual patients:

We deal very much with rehabilitation, to whatever the patient wants to be rehabilitated to. One of our patients considers himself really rehabilitated, and that is because he is able to comfortably walk back and forth to his car if he wants to go out and get a pack of cigarettes. And he can stay awake for the soap operas. And to him, that is rehabilitated. I mean, he is so happy, he said, "You know, I am just the way I want to be."

Several caregivers discussed the concept of acceptance and the need to assist the patients to accept their condition in order to promote positive quality of life. A dialysis unit head nurse described it this way:

I think that they [patients] do reach a level where quality of life is very good, because they make it that way. They accept somehow, that thing about "oh, you'll get used to it" or "everything will be all right in the morning"—that's what my father used to tell me. And somehow it happens, even though, you know, you weather all sorts of storms, somehow you come out of it a little dryer. You say, "Well, things aren't that bad, and if I can just get through this period," you know, so it's the crisis situations that you really have to get the patients through, to really appreciate the quality of life. And not all of them do that. They do not accept it, but—with the pretends or the make-believes that they go through—they still somehow have a fairly decent life, even with all of the denial.

One unit staff nurse related age variations to difference in quality of life for her patients:

In a way I think it's easier for the older patients. Their life is partly used up. The younger ones [dialysis patients] have difficulty accepting. We just had this young 25-year-old man who got started. He called himself—what is the word—"handicapped." He said, "Now, because of this I have to quit my job, stop exercising, doing all the things I like."

Several caregivers discussed economic differentials as being related rather indirectly to quality of life. One noted, "If they [the patients] are financially secure, the quality should be very, very good. If they are living in a very depressed economic area, well they have all the problems that everybody else does, plus their renal disease on top of it." Another caregiver added,

I think that socioeconomic level definitely has an impact on quality of life. And I think persons in higher socioeconomic levels have a better developed coping; maybe that's a blatant generalization. But I think it can perhaps be true also. It is harder on an upper socioeconomic status person because more is taken away from them, whereas some patients who maybe were on welfare didn't do a whole lot before, and so their life hasn't changed all that much. Someone who is very young and dynamic and heading for a career, this is like a monkeywrench thrown in it.

A hemodialysis unit head nurse summarized her perception of the chronic dialysis patient's life this way:

My own perception of it is that I think we could learn a lesson from these patients. I think that for the most part, by and large, they are really great people. Because if I were presented with chronic renal disease today, I don't know that I'd want to survive, I really don't. I think that they are survivors. I think they're very strong people, and I think for that they should be given credit. But I think that their lifestyles have been altered enormously. I'm always a little depressed when I see someone whom I haven't seen in a while, who perhaps we've started and sent out, and isn't doing so well now. It is just upsetting to me to see them. I never really think that they're doing all that well—even the people that are rehabilitated, that go back to work, they say they get tired. You know, I'm just not so sure when you have to devote 12 hours a week to some treatment, that you can expect to have a fine life. I think some people do as well as they can, and for that they should be given credit. I guess we would all like things to be the way they were at some point in time, but I'm not so sure that I could say that I think the quality of their life is something I would want. I think it's a little unnatural.

UNCERTAINTY OF THE FUTURE

It has been suggested that rehabilitation or "the ability to return to a nearly normal way of life" is the "ultimate goal" of the dialysis patient.[34] Even when this goal has been achieved, however, the patients must live with many uncertainties relative to their physical condition. Hemodialysis patients are very aware of the potential dangers of each treatment session— risks such as air embolus, hepatitis, and severe hypotension. They are also aware of the risks related to missing even a single treatment—elevated chemistries, including dangerously high potassium (K^+), which might result in cardiac arrhythmias; fluid overload with resultant pulmonary edema; and hypertension. Patients are conscious that their access mechanisms, be it shunt, fistula, or graft, must be continuously cared for and kept functioning in order to provide the necessary link to the life-giving hemodialysis treatment procedure. Finally, dialysis patients are ever aware that they have a serious, potentially life-threatening physical condition, end-stage renal disease, which will gradually bring about deterioration of a number of their body systems.

The concept "future," as well as the phrase "long-term hemodialysis patient," is relative for the dialysis patient population. To date it is reported that some patients have survived up to 15 or 20 years (perhaps a few even longer) on hemodialysis, but the majority of "long-term" patients, such as those interviewed in the present research, have a hemodialysis history of only about 9–12 years. Thus, it is difficult for the professional medical community, much less the patients themselves, to establish a plan for any more distant concept of future. There have been a number of scientific breakthroughs during the past two decades for ESRD patients, most notably the possible alternative of either related living or cadaver kidney transplantation. This procedure, however, is not without its own risks and uncertainties. More recently CAPD and CCPD are being advocated and elected as treatment of choice for certain patients. With these newer modalities, also, long-term prognosis is uncertain.

Interestingly, while hemodialysis caregiver respondents in the study frequently commented on and discussed the uncertain future of their patients, the patients themselves did not focus much upon the topic. On the contrary, patients appeared to be much more oriented to the present, to living each day as it came, grateful for another 24 hours of survival. One young male patient, 35-year-old, who was not a long-term survivor in the study (the patient expired after approximately three years on dialysis), commented that he became less concerned about the future after learning that he had ESRD than before the onset of his illness. He reported that prior to becoming ill he had been working on a doctoral degree, working full-time, and was totally immersed in achieving his future goals. "Now," he noted, "I have really changed. I have become much more tolerant than I was. Now also I notice the beauty of nature, the types of people. Before I was going around like a blind person—now I see a red sky at sunset. I have time to go to concerts. Before, I was so much career-oriented. I don't have to go up on the social ladder now. I enjoy life more deeply than I ever did before."

A number of other patients expressed similar attitudes toward their "before" and "after" dialysis existence, particularly noting positive changes in family relationships.

It was found, however, that several study respondents who had been "on the machine" for longer periods of time (9, 10, 11 years) did begin to express some concern about their length of time on hemodialysis and the related physical side effects, such as neuropathies. One patient, a 10-year dialysis veteran, commented that he was "wearing down" on hemodialysis. Another patient decided to try CAPD for a similar reason.

A male patient in the study discussed the uncertainty of his survival indirectly, commenting upon his worry concerning his children's college educations; and several others mentioned similar future-oriented family concerns, should they not survive. As a rule, however, the topic of a patient's own demise was generally avoided. It is suggested that because ESRD patients live so intimately with the awareness of death, some "healthy" denial must be initiated in order that their ordinary day-to-day functioning be carried out; that is, some denial of their uncertain future must be employed in order to foster within the patients' the necessary "courage to survive."

REFERENCES

1. Brundage DJ: Nursing Management of Renal Problems. St. Louis, CV Mosby, 1976
2. Lancaster LE, Pierce P: Total body manifestation of end-stage renal disease and related medical and nursing management, in Lancaster LE (Ed): The Patient with End-Stage Renal Disease. New York, Wiley, 1979, pp 1–60, p 1
3. Czaczkes JW, Kaplan De-Nour A: Chronic Hemodialysis as a Way of Life. New York, Brunner/Mazel, 1978, pp 126–127
4. Statistical Analysis on Patients in Reporting U.S. Dialysis Centers. (Paper prepared for the Artificial Kidney–Chronic Uremia Program, National Institutes of Health, Bethesda, Md), North Carolina, Research Triangle Institute, July 1, 1975
5. O'Brien ME: Hemodialysis and Effective Social Environment: Some Social and Social–Psychological Correlates of the Treatment for Chronic Renal Failure. Unpublished doctoral dissertation, The Catholic University of America, Washington, D.C., 1976, p 62
6. O'Brien ME: Hemodialysis and Effective Social Environment: Some Social and Social–Psychological Correlates of the Treatment for Chronic Renal Failure. Unpublished doctoral dissertation, The Catholic University of America, Washington, D.C., 1976, pp 66–67
7. O'Brien ME: Hemodialysis and Effective Social Environment: Some Social and Social–Psychological Correlates of the Treatment for Chronic Renal Failure. Unpublished doctoral dissertation, The Catholic University of America, Washington, D.C., 1976, pp 60–77
8. Eccard M: Psychosocial aspects of end-stage renal disease, in Lancaster LE (Ed): The Patient with End-Stage Renal Disease. New York, Wiley, 1979, pp 61–68, p 61
9. Kaplan De-Nour A, Shaltiel J, Czaczkes JW: Emotional reactions of patients on chronic hemodialysis. Psychosom Med 30:521–533, 1968, p 526
10. Czaczkes JW, Kaplan De-Nour A: Chronic Hemodialysis as a Way of Life. New York, Brunner/Mazel, 1978 p 112
11. Schlesinger L: Disruptions in the personal social system resulting from traumatic disability. J Health Human Behav 6:91–98, 1965, p 91

12. Schlesinger L: Disruptions in the personal social system resulting from traumatic disability. J Health Human Behav 6:91–98, 1965, pp 97–98

13. Sorensen E: Group therapy in a community hospital dialysis unit. JAMA 221:899–901, 1972, p 899

14. Nordan R, Ostendorf R, Naughton JP: Return to the land of the living: An approach to the problem of chronic hemodialysis. Pediatrics 48:939–945, 1971, p 939

15. Kaplan De-Nour A, Czaczkes JW: Personality and adjustment to chronic hemodialysis, in Levy NB (ed): Living or Dying, Adaptation to Hemodialysis. Springfield, Charles C Thomas, 1974, pp 102–126, p 104

16. Brundage DJ: Nursing Management of Renal Problems. St. Louis, CV Mosby, 1976, p 133

17. Czaczkes JW, Kaplan De-Nour A: Chronic Hemodialysis as a Way of Life. New York, Brunner/Mazel, 1978, p 140

18. Goffman E: Stigman, Notes on the Management of a Spoiled Identity. New Jersey, Prentice Hall, 1963, pp 1–4

19. Hansen GL: Psycho-social problems related to chronic hemodialysis, in Hansen GL (Ed): Caring for Patients with Chronic Renal Disease. Philadelphia, Lippincott, 1972, pp 68–73, p 70

20. Dickerson Z: Stress factors in hemodialysis. Nephrology Nurse 1:19–21, 1980, p 20

21. Abram HS: The psychiatrist, the treatment of chronic renal failure and the prolongation of life, I. Am J Psychiatry 124:1351–1357, 1968

22. Abram HS: The psychiatrist, the treatment of chronic renal failure and the prolongation of life, II. Am J Psychiatry 126:157–167, 1968

23. Abram HS: The psychiatrist, the treatment of chronic renal failure and the prolongation of life, III. Am J Psychiatry 128:1534–1539, 1972

24. Goldberg RT: Vocational rehabilitation of patients on long-term hemodialysis. Arch Phys Med Rehabil 55:59–65, 1974, pp 63–64

25. King S: Social–psychological factors in illness, in Levine S, Reeder L, Freeman H (Eds): The Handbook of Medical Sociology. Englewood Cliffs, New Jersey, Prentice-Hall, Inc., 1972, pp 129–147, p 138

26. Bailcy G: Psychosocial aspects of hemodialysis, in Bailey G (Ed): Hemodialysis: Principles and Practice. New York: Academic Press, 1972, pp 430–440, p 432

27. Hampers C, Schupak E: Long-Term Hemodialysis. New York, Grune and Stratton, 1967, p 147

28. Litman TJ: The influence of self-conception and life orientation factors in the rehabilitation of the orthopedically disabled. J Health Human Behav 6:249–257, 1965, p 255

29. Shambaugh P, Kanter S: Spouses under stress: Group meetings with spouses of patients on hemodialysis. Am J Psychiatry 125:100–108, 1969, p 102

30. Sorensen E: Group therapy in a community hospital dialysis unit. JAMA 221:899–901, 1972, p 900

31. Cummings J: Hemodialysis: feelings, facts, fantasies—the pressures and how patients respond. Am J Nurs 70:70–76, pp 71–72

32. Parsons T, Fox R: Illness, therapy and the modern urban family. Journal of Social Issues 8:31–44, 1952, p 32

33. Coe R: The Sociology of Medicine. New York, McGraw-Hill, 1970, p 74
34. Holcomb JI: Social functioning of the artificial kidney patient. Soc Sci Med 7:109–119, 1973, p 109
35. Crammond WA, Knight PR, Lawrence JR: The psychiatric contribution to a renal unit undertaking chronic hemodialysis and real homotransplantation. Journal of Psychiatry 113:1201–1212, 1967, p 1203
36. Mechanic D: Religion, religiosity and illness behavior. Human Organization 22:203–208, 1963
37. Safilios-Rothschild C: The Sociology and Social Psychology of Disability and Rehabilitation. New York: Random House, 1970, pp 119–120
38. Schowalter J, Ferholt J, Mann N: The adolescent patient's decision to die. Pediatrics 55:97–103, 1973, p 100
39. Ulrich B: The psychological adaptation of end stage renal disease: A review and a proposed new model. Nephrology Nurse, 3,3:48–58, 1980, p 49
40. Cummings JW: Hemodialysis: feelings, facts, fantasies—the pressures and how patients respond. Am J Nurs 70:70–76, 1970, pp 71–72
41. Crammond WA: Psychological aspects of the management of chronic renal failure. Br Med J 1:539–543, 1968, p 540
42. Levy NB: Sexual adjustment to maintenance hemodialysis and renal transplantation: National survey by questionnaire: preliminary report, in Levy NB (Ed): Living or Dying, Adaptation to Hemodialysis. Springfield, Ill, Charles C. Thomas, 1974, pp 127–140, p 130
43. Levy NB: Sexual adjustment to maintenance hemodialysis and renal transplantation: National survey by questionnaire: Preliminary report, in Levy NB (Ed): Living or Dying, Adaption to Hemodialysis. Springfield, Ill, Charles C. Thomas, 1974, pp 127–140, p 139
44. Collins JL: Multidisciplinary consultation to a renal dialysis-kidney transplantation unit. Unpublished paper, Walter Reed Army Institute of Research, May 1974, pp 1–12, p 8
45. Milne JF, Golden JS, Fibus L: Sexual dysfunction in renal failure: A survey of chronic hemodialysis patients. Int J Psychiatry in Med 8:335–345, 1978, p 343
46. O'Brien ME: Hemodialysis and Effective Social Environment: Some Social and Social–Psychological Correlates of the Treatment for Chronic Renal Failure. Unpublished doctoral dissertation, The Catholic University of America, Washington, D.C., 1976, pp 106–107
47. Finkelstein SH— Finkelstein FO, Steele TE: Dialysis marriages. J AANNT 1:13–16, 1974, p 14
48. Alexander RA, MacElveen PM: Are we assesing needs of home dialysis partners? J AANNT 4 (suppl):23–28, 1977
49. O'Brien ME: Hemodialysis and Effective Social Environment: Some Social and Social–Psychological Correlates of the Treatment for Chronic Renal Failure. Unpublished doctoral dissertation, The Catholic University of America, Washington, D.C., 1976, p 112
50. Sorensen E: Group therapy in a community hospital dialysis unit. JAMA 221:899–901, 1972, p 900
51. Bailey GL: Psychosocial aspects of hemodialysis, in Bailey GL (Ed): Hemodialysis: Principles and Practice. New York, Academic Press, 1972, pp 430–440, p 434

52. O'Brien ME: Effective social environment and hemodialysis adaptation. J Health Soc Behav 21:360–370, 1980, pp 364–365

53. Steckel JB: The use of positive reinforcement in order to increase patient compliance. J AANNT 1:39–41, 1974

54. Roper E, Raulston A, Cramer D: Attitudinal barriers in dialysis communication. J AANNT 4:179–198, 1977

55. Gelman GB: Peer counseling: Patients helping patients. Nephrology Nurse, 1,4:19–21, 1979, p 19

56. Mulkerne D, Tucker CM: Peer facilitation training with hemodialysis outpatients. Nephrology Nurse, 2:58–60, 1980, p 59

57. Broncatello KF: Auger in action: Application of the model, Advances in Nursing Science, 2:13–23, 1981, p 19

58. Freidson E: The Profession of medicine. New York, Dodd, Mead and Co., 1971, p 58

59. Susser MN, Watson W: Sociology in Medicine. London, Oxford University Press, 1971, p 60

60. Mechanic D: Religion, religiosity and illness behavior. Human Organization 22:202–207, 1963

61. Beard BH: Fear of life and fear of death. Arch Gen Psychiatry 21:373–783, 1969

62. Norton CE: Attitudes toward living and dying in patients on chronic hemodialysis. Annals NY Acad Sci 164:720:732, 1969

63. Levy NB: Living or Dying—Adaptation to Hemodialysis. Springfield, Ill., Charles C Thomas, 1974

64. O'Brien ME: Effective social environment and hemodialysis adaptation—a panel analysis. J Health Soc Behav 21:360–370, 1980

65. Foster FG, Cohn GL, McKegney FP: Psychobiologic factors and individual survival on chronic renal hemodialysis: A two year follow-up: Part I. Psychosom Med 35:60–67, 1973, p 67

66. Barry AM: Nursing actions perceived by patients adjusting to chronic renal disease and intermittent hemodialysis. Unpublished master's thesis, The Catholic University of America, Washington, D.C., 1969

67. Calsyn DA, Sherrard DJ, Freeman CW, et al: Vocational adjustment, psychological assessment and survival on hemodialysis. Trans Am Soc Artif Intern Organs 24:125–126, 1978, p 125

68. Goldberg RT: Vocational rehabilitation of patients on long-term hemodialysis. Arch Phys Med Rehabil 55:60–65, 1974, p 64

69. Jackle MJ: Life satisfaction and kidney dialysis. Nurs Forum 13, 4:361–370, 1974, p 362

70. Jackle MJ: Life satisfaction and kidney dialysis. Nurs Forum 13:361–360, 1974, p 364

71. Lauer RH: Social Problems and Quality of Life. Dubuque, Iowa, Wm C Brown, 1978, p 166

72. Gutch CF, Stoner MH: Review of Hemodialysis for Nurse and Dialysis Personnel. St. Louis, CV Mosby, 1975, p 185

73. Kaplan De-Nour A, Czaczkes JW: Professinal team opinion and personal bias— a study of a chronic hemodialysis unit team. J Chronic Disease 24:533–541, 1971, p 534

74. Kagan LW: Renal Disease. New York, McGraw-Hill, 1979, p 188

—————3—————
The Families

We are made by relationships with other people.

Carlo Carretto
Summoned by Love

At one time the concept of family was more easily defined within American society. A classic definition was that of Murdock in 1949:

The family is a social group characterized by common residence, economic cooperation, and reproduction; it includes adults of both sexes, at least two of whom maintain a socially approved sexual relationship, and one or more children, own or adopted, of the sexually co-habiting pair.[1]

Murdock's definition encompasses the concepts of both conjugal or nuclear family, i.e., the married or socially approved pair with or without offspring, and the consanguine or extended family, i.e., that group founded upon "the blood relationship of a large number of kinspersons."[2] In general, the functions of such families have been thought to consist of activities such as "the regulation of sex and reproduction; the care and socialization of children; economic co-operation; intimacy and companionship; and maintenance of the social stratification system."[3] In contemporary society, however, the term "family" is also being used to describe a variety of groupings of nonkinship or legally sanctioned relationships. One now may participate in a "communal" family (nonreligiously oriented) of persons living together for reasons of economic benefit or mutual support. In certain cultural groups persons with no blood ties or nuclear family linkages will be referred to as "play" family, e.g., play brother, play aunt. These individuals may or may not reside in the same household. Close friends may be labelled "family" by the single, nonparent; and, on occasion, even those persons with whom one has primarily instrumental relationships, such as service providers, will be labelled "family" when the interaction includes an affective element of mutual caring and concern.

In considering the subject of "family" in relation to the hemodialysis patient, several types of supportive relationships have been included. The 26 individuals with relationships with dialysis patients who provided data for the present study, included spouses, parents, children, kin (e.g., cousin), and friends. The profile of the respondent group is as follows: 12 spouses, 4 friends, 3 mothers, 1 cousin, 5 daughters and 1 son.

SUPPORT OF SIGNIFICANT OTHERS

There has been much discussion in the literature focusing upon the importance of family support for the adaptation of the chronic dialysis patient. It has been suggested that,

> Because of the complex psychosocial problems associated with any chronic disease, and particularly because of the unique condition of the chronic hemodialysis patient who lives dependent upon a machine, interaction with family, or other significant persons is important in helping the patient cope with necessary modifications in behavior, and accept both his condition and the prescribed treatment regimen.[4]

Kossoris asserts that, "The patient with renal disease is very dependent on his family, and as his disease progresses, so does his involvement with and dependence upon family members."[5] Researchers at the University of Washington have pointed out that for those patients whose adjustment to maintenance hemodialysis appeared most successful, a significant finding was the presence of well-developed emotional support systems on the part of the family members.[6]

King takes the position that the potential for stress in the relationship between the ill person and his or her family is more likely to be found in the nuclear family than in the extended kinship group.[7] Litwak and Szelenyi argue that, while the nuclear family meets those needs traditionally considered to be in the realm of the primary group, a key deficit is "its lack of human resources."[8] Where a patient's family consists only of a spouse or perhaps a spouse and a small child, the heavy burden of decision making, as well as emotional support, may well fall on the shoulders of one individual. Where there is a larger extended family system, members of the kinship group frequently take over responsibilities and provide back-up support for members of the nuclear family in times of stress and crisis such as the occurrence of serious illness. Parsons and Fox assert that any type of family has difficulty in maintaining a healthy balance between the "supportive and dependent aspects of illness, and the discipline necessary for compliance with the treatment regimen."[9]

Theodor Litman notes that ever since Richardson's classic treatise, *Patients Have Families,*[10] social scientists have studied the role of the family in regard to adjustment to and response to illness.[11] He cites Sussman as having undertaken an extensive review of the literature in order to establish the acceptability of applying "certain measures of family solidarity to the study of chronic illness and disability."[12] A particular problem often encountered in chronic illness, however, is the difficulty that family members have in providing the ill loved one with the necessary concern and emotional support without becoming oversolicitous and overprotective, placing the patient in a dependent and undisciplined position. Litman reported a strong correlation between family reinforcement and satisfactory convalescence in the orthopedically disabled, but suggested that in some cases, "close family ties may actually serve to deter rehabilitation."[13] This latter comment is supported by Czaczkes and Kaplan De-Nour, who argue that family support may not be a necessity in order to help all patients adjust to chronic dialysis. They suggest, rather, "that the influence of family reactions on the patient's adjustment depends not so much upon the reactions per se but more on the personality traits and the emotional needs of the patient."[14] They add that the type of family reaction that may help one patient to adjust may cause problems for another patient.

In the present research, one aspect of family support that was discussed by patient, family members, and caregivers as well, was that related to family members' expectations regarding patient behavior, both for adaptation to the hemodialysis treatment regimen and general social functioning in terms of work, recreation, and group interaction.

Overall, the majority of family members interviewed reported that they had tried to avoid "overprotectiveness" or treating the dialysis patient as an invalid, but did try to be sympathetic when he or she was feeling badly. The son of a 12-year hemodialysis "veteran" commented that he felt the family had helped his father cope through not making "an issue" of his dialysis. He said, "You see, we never saw the illness as anything special or anything earth-shattering. It was just something to deal with. And I think that probably helped in the long run because that's how we deal with things in general."

This perception was confirmed by the patient himself, who reported that, although not always able to be the "breadwinner," he had been able to successfully fulfill his "fathering" role responsibilities throughout the course of his illness.

The wife of a 10-year dialysis patient noted the importance of avoiding "pity" in her interactions with her husband, an attitude that could prove demeaning and emasculating to a male patient. She described her feelings and behavior this way:

He is a fantastic person. He has not let any of this get him down. I treat him nice but I don't pity him—I don't give him sympathy. He says, "I feel bad, I'm sick," and I say, "Well, you know there are some people who can't get out of bed this morning and you can get up and get your clothes on, you have a brain, you can think." I really don't pity him, so I think it keeps him going. Sometimes I have to get him involved, if I see he's getting a little depressed, I'll find a little something, a little argument to pick with him—nothing that's going to upset him, but just enough to get him started and defend himself and then he gets all worked up. Not enough to run his pressure up but I don't let him get depressed.

The daughter of a long-term female dialysis patient described her expectations and support of her mother this way:

Dependence is a big problem for my mother. She's tired of relying on other people to do things for her. She was always one of those spotless housekeepers. She's really had to change that. You have to expect some disablement. You can't possibly expect them to do what they used to. But a lot of patients go very downhill if everything's done for them. I think the reason my mother has lasted so long on dialysis is that I haven't done everything for her. Of course, sometimes you have to, if they really don't feel well. But I make her do for herself. I have forced her to do even when she didn't think she could and then she would feel good about it and say, "Look what I did!" You don't treat them like a baby either or they get very resentful. And you don't treat them like they're sick. I mean, you don't have them do heavy vacuuming and stairs and windows and all that, but just light work—what they can do.

Finally, a close female friend and housemate of a 12-year female dialysis patient described her attitude and behavior in regard to promoting independence on the part of the patient:

Some things we don't do together because I insist she do them by herself, like shopping. She is capable of going out alone, normally. When she decides to go out I know there are people there who are going to look out for her; I know that she is not so helpless that she will not be able to tell someone to call home and let me know what's going on. It's all an exercise in independence. A lot of times she wants to go when I don't, so I say, "Go, I'll stay right here at home and be perfectly all right." It's really my outlet—she's gone and I don't have anything to worry about and I don't worry when she's away, because I know she's gonna be all right and so far, so good. I have not been wrong. Besides, I know she can always call me and I can go and bring her home.

This respondent also described and analyzed her "way" of caring for her friend. She noted that one should give "good caring, not going overboard," and added,

You'll find that is one of the problems. People can't be doted on constantly. If someone did it to me then I'd probably get worse, depressed about this person raining his affections on me all the time, never letting me do what I'm capable of doing. The key point is letting a person risk for the sake of their independence—not telling them that they can't do. Yes, they can, and they need someone to tell them [they can] in order for them to be successful. I know what she can and cannot do. Those things that she cannot do, I don't press. There's not very many things she can't do. I don't like to see her in her wheelchair a lot; I certainly take the wheels off. She can walk and I know she can walk and I tell her, "God gave you two legs and you can still use them," but I also know when she needs that chair, when she comes in from the machine and she's so weak, and then it's okay. There are very few things she can't do and even when there is something, we sit down and try to work it out so maybe she can do it now, with another approach. And if we exhaust three or four different methods and it doesn't work out, and we're back to square one, maybe we still try another plan. If a person is as close to you as she is, it's something that you feel. As soon as she pulls herself together then I know that, too. But even if she looks well, I know when she's not—that's how close we are. Maybe that's the test of knowing whether or not you really care. It's being able to feel that. Feelings like that don't lie.

Most married study patients reported receiving much support and concern over time from their spouses, and several attributed their continued survival to such care and encouragement. A male patient described his wife as "being right in it with him" over the years, and commented, "Without her there's just no way I could have made it this long." A female patient described her spouse's support in a similar vein: "He really took over when I got sick and just did everything. I don't think I could have ever handled this thing alone."

Some of the dialysis patient study respondents reported having no close family or kin to provide help or support. A few patients mentioned having friends who helped them on occasion, and one gentleman described his situation this way:

Well, the only people who ever helped me out were what you would call winos. When I would come home from the unit and couldn't get around much, they would clean up and go to the store for me and wouldn't ask for anything in return. They came by every day to check and see if I needed anything.

Dialysis patient support systems varied a great deal, as was validated by a hemodialysis unit social worker who noted,

We do get a significant number of patients who are just loners out in the community. They really don't have family or community or friends' support. Or else, maybe they lose them. Somewhere along the line they have lost them. By the time they get to us, they are kind of on their own. But we do have other families who are very, very much involved and who are willing to do whatever they can to help the person.

The Cab Driver

A rather unusual relationship, which involved a range of interactional behaviors and types of support for certain dialysis patients, was that of the patient and cab driver. In Chapter 5, one female patient is described who refused to change her dialysis treatment from a center quite a distance from her home to one quite close "because of her cab driver." This respondent commented that for years she had depended upon one particular cab driver who performed many small services for her, especially during the period when she was more seriously limited physically. She said, "I couldn't have made it without him. He came up to the apartment, helped me with my coat, got me in the wheelchair. And then he'd drive me to the unit, get some food, get me some ice and get me to my machine. We're good friends." Similar types of cab-driver–patient relationships were reported by other respondents and caregivers. One female patient stated, "My cab driver really knows me. She lets me stop and get food [on the way to treatment] and sometimes she buys it for me. She said she was going to come and take me to dinner when I got my appetite back." Another patient noted,

My cab driver, he picks up four of us for the mid-day shift and carries us to the unit. He's the steady cab driver—if his cab breaks down, he'll call another cab. He'll see to his regular customers, us getting here on time. I got his phone number and he has mine. If I take sick, I call him and if he can't make it, he calls me. That's the way he does all his patients. He keeps those close who need a helping hand. When we get one cab driver like that, we try to stick together. He sometimes gets us lunch at the carry-out, too.

A patient commented on other special services of the drivers, pointing out the generosity with which some of the dialysis patients were treated:

Those cab drivers are really nice. They bring in extra ice for the older patients. Sometimes they buy lunch or dinner. If people don't have the money they even buy it for them.

Another respondent reported, "One time this winter they [the cab drivers] went about four months without gettin' paid [drivers are reimbursed under the federally funded ESRD program] but they still kept pickin' people up." Several caregivers also commented on the assistance that patients received from their cab drivers and the dependence on their services. One shift supervisor described her feelings toward the drivers this way:

We're taking up money to give [to] cab drivers who take care of the patients as a Christmas present; you know, they've done such a good job. The cab drivers get very involved in some of these patients. I don't know if you know F.—he's a patient here. He's an older gentleman who's not particularly "with it" from a mental status, you know, but his wife and his daughter take excellent care of him. He's very clean, very meticulous. He can't walk and he can't talk. He just kind of sits there. And one of the nurses went over to give him a shot . . . and she wanted to turn him over and take his pants down, and this cab driver went wild. He said, "He has his right to dignity. He has feelings. . . . Don't you dare!" They really care.

The supervisor also described another driver who frequently brought patients to her unit: "He's very good. He'll call me up and say, 'So and so, I can't get her up. What do you want me to do?' We have a very strong working relationship." She continued,

> Whatever I ask the drivers to do, they'll do for me. Patients I've needed taken to the hospital on an emergency basis and they weren't needing an ambulance, they say, "Well, I'll take them for you." They're good and they do give the patients good care. They go out and get their breakfasts for them. They drop the patients off and go back out and make a trip to the store.

Finally, a cab driver who had been transporting dialysis patients to and from their treatment sessions for approximately 8 to 9 years, put it this way:

> You get to know people for 5 or 6 years and it's more than just a business relationship. I do some things for the patients. Some people don't really have anybody to depend on, to do for them. It's just something that you do. After being with somebody [driving them to and from the dialysis unit] for 7 or 8 years, it's hard not to, especially when you know they need it. Actually you befriend somebody, it's a friendship relationship—it's just what you do.

The relationship of the hemodialysis patient and the cab driver takes on special importance because of the survival-related meaning assigned to patient transportation to the dialysis unit, i.e., the life-continuing implications of the treatment sessions. These "life and death" implications associated with an instrumental type, service-oriented relationship, i.e., public transportation, appear to have provoked the initiation of an affective dimension in the participants' interaction. Cab drivers were often observed to escort patients into the unit, settle them in their dialysis chairs, provide them with food and drink to be enjoyed during the treatment session and, in fact, behave quite protectively toward their "patients." An element of mutual respect, caring, and concern was evidenced in the relationships.

In evaluating the expectations of the family and friend respondent groups, it was found, as the above examples suggest, that most family members and friends of long-term patients viewed their role as consisting of support, caring, and occasionally taking over certain duties or role responsibilities usually allocated to the dialysis patient. The majority of respondents, however, did expect their patient family member to be as self-sufficient as possible and to perform to the level their physical conditions allowed at any particular time. Some family members commented that early on in the patient's adaptation, less was expected by the family, but as time wore on and "everyone" got more used to the dialysis routine, a kind of normalization or "routinization" of expectations and behavior began to occur on the part of both patient and family.

IMPACT OF ESRD AND DIALYSIS ON THE FAMILY

Family reactions to illness in general, and to end-stage renal disease and the hemodialysis treatment regimen in particular, have frequently been discussed in the literature. Duff and Hollingshead observe that in facing any illness family members either draw together to cope with the problem or draw back and attempt to avoid any personal involvement.[15] They label this drawing together "centripetal" activity, resulting in the mobilization of resources to meet the imminent threat to the family's integrity; they view the pulling back of family members into individual isolation as "centrifugal" force. In their study of the impact of illness and hospitalization, Duff and Hollingshead measured centripetal and centrifugal tendency by estimating degree of empathy between the family members and the patient. Empathy was described as "the ability to understand and identify with another person,"[15] and was viewed as generally important in family relations, and even more so during illness.

In regard to the plight of the chronic hemodialysis patient and his or her family group, many unique sources of social and social–psychological stress may be found. Hampers and Schupak argue that "there is perhaps no situation so stressful to patients and their families as chronic hemodialysis,"[16] as major adjustments in thinking and living must be made. Interpersonal relationships between patients and their families and friends may be altered notably, and often the individual is faced with a considerable degree of social isolation. Viederman points out that one contributing factor to such isolation is that the patient's continued existence is contingent on being attached to a machine for 4 to 6 hours several times a week, a machine from which the patient "may never stray too far."[17]

Role Reversal

In discussing illness in general, King maintains that if decision making in a family is a joint action shared by husband and wife, the process is usually not seriously hindered by the illness of one member.[18] The reverse, however, is true if decision making for the family group is carried out primarily by one member and that member should fall ill. In this case a situation often arises to force the weaker member of the pair to take on a role and responsibilities for which he or she is largely unprepared. These new role responsibilities may then result in anxieties that the role incumbent finds difficult to conceal from the ill partner.

In the case of hemodialysis patients, Sorensen found that "a frequent source of difficulty lies with the spouses who live under the responsibility of caring for the patient, and this generates much tension."[19] Similarly, Bailey determined that when a wife learned to carry out the dialysis treatment for her spouse, role-reversal began to occur, as the former protector and bread-winner now became dependent upon his wife for his very continuance of life.[20] Jonathan Cummings, in a study of male dialysis patients at a metro-politan veterans hospital, found that social role disturbance was clearly ver-ified in his patient population and noted that particularly susceptible to the strain of dialysis were the roles of "breadwinner, disciplinarian and decision-maker."[21] In discussing this concept of role reversal, Brundage added that "repeated stresses, uncertain future and altered marital relationships are threats to family functioning."[22] Brinker and Lichtenstein focus on the dual role imposed upon the healthy spouse of a dialysis patient; they note: "Role change is common in dialysis families such that spouses often take on the role responsibilities of the sick partner while maintaining their own role. This reduces rest and leisure and lowers physical reserve."[23]

Role reversal and role reorganization were commented on either di-rectly or indirectly by a number of family-member study respondents. The wife of a dialysis patient stated that she "has had to do most of the disciplining [of the children] since her husband's illness"; and had basically assumed all of the parental responsibilities. She added that she had really had to take over the running of the household, and commented: "I do the financial things, pay the bills and all that. He just doesn't seem to want to get involved now." Another wife noted that she did not want to be involved with home dialysis precisely because of having to take on other family responsibilities. Her husband validated the fact, adding, "The home thing—she doesn't want to get involved with that. She has enough to do without taking care of me on the machine." A dialysis patient's husband commented on his perception of role reorganization, stating that he had basically taken over all of the household cleaning chores for a period of time, as well as carrying his own "money-earning" work.

A female housemate of a long-term female dialysis patient described how she handled the formerly-shared household roles:

Well, she's able, but I'm more capable, so I'll take over the majority of things—it gives her more time to do other things. I do all the cleaning and cooking. Once in a while she'll treat me and that's good because it's not good for her to lose track of all the things she used to do. I'll force her to do things too. Sometimes I'll blatantly refuse. I'll say, "I will not"—and she'll get up and she'll do it and I'll say, "You can do anything you want to do," and I make her do things, and after it's all over with, she's happy with herself and I'm happy. It takes a bit of brow-beating sometimes, and I put my foot down and that's exactly what I mean and she knows it. Even our friends are really amazed at our relationship; they don't understand. They'll say, "Why do you do that?" or "How?" and I say, "Because she loves me and I love her." That's the whole core of our relationship, because we love each other.

In alluding to a parent–child role-reversal situation, the teenage daughter of a female dialysis patient reported on her assumption of parental and housekeeping responsibilities for several younger sisters. She said, "I took on a lot of responsibility, but it's not too hard. I take care of my mother and I do cooking and washing, and when there's something important like a problem at school about one of my sisters, then I have to go." She added, "I'm trying to find a job, too. I've been working off and on, but it's hard to get a babysitter so often."

Family Interaction

Some dialysis patient respondents in the present study reported that their illness condition had, in fact, exerted a "centrifugal" kind of force, causing family members to pull away into their own "worlds" and activities. One older patient admitted that once he got over the initial adjustment to ESRD and dialysis, his family seemed to "lose interest" in his condition and no longer gave him the attention that he felt would have been appropriate; and another younger male patient noted, "Your friends don't come around once you can't keep up with them." The majority of patients, however, seemed to perceive a "centripetal" effect from the illness, causing family members to grow closer together. A female patient reported, "I have gotten much closer to my husband and children [since ESRD]. I really don't have many friends any more. There's no time to see them. But my husband has become much more supportive than ever before." The wife of a patient stated, "Now we are very close in the family. My son helps take care of him [the patient]. It's good we are all together. We help each other." Another female dialysis patient's cousin noted what she described as one "good thing" about the illness condition: "Well, she and I are closer together. And she has a much better relationship with her mother, too." Finally, a "mother" dialysis patient, an 11-year veteran, reported that the treatment regimen had been a "mixed blessing." She commented that it has brought her family together to help and support her, but it had also caused them much worry over the years. She said, "I think dialysis has changed everybody's life in the family. I don't know if for the better or worse. Sometimes worse, because you could have done *more* in your own life and for them."

Family Social Activities

In general, most family members of study dialysis patients admitted to restrictions in their social activities related to ESRD and dialysis. The wife of one long-term dialysis patient commented,

> I figure we are restricted because I'm a person who loves to go and loves to do things and I realize that he doesn't feel like it a lot of times. We go out and we do things with the kids, but not as many things as we would do if things were different. For instance, before he was on the Monday, Wednesday, Friday shifts, we couldn't go out of town on weekends because he would have to go on the machine on Saturday evening. Now he wants to go on a cruise this summer, but I don't know what's going to happen because, well, we go day to day.

Restriction in the family's ability to travel was reported by several family members. The son of a dialysis patient commented that the family had had some extensive travel plans, but noted,

> When he [his father] got sick all those plans were cancelled and we hadn't done that much traveling. We were sort of saving it. And I remember feeling a little bit upset about it, but it didn't really matter, I mean, at least to me. You know, it was just accepted, and we weren't going to go without him.

The respondent added that he was somewhat worried about his father because of the restriction, observing, "I am concerned about my dad confining himself because he is on dialysis," but admitted that it may be because of an unpleasant past experience his father had had as a "visitor" in a dialysis unit in another state. He commented, "I can understand why my dad won't travel, because he did have a bad experience when we were on vacation and I know he doesn't want to go through that again." A dialysis patient's spouse noted also that their social life was restricted following her husband's bout with hepatitis. She reported, "Before he was very active. He liked to swim and play tennis. Now he's too tired and he lost his muscles after the blood infection. He lost all his strength." The mother of a female patient supported the concept of restricted activities with this comment: "Dialysis has really stopped a lot of things, because we could go places but she has to be back at a certain time for dialysis. We used to really go a lot more." The wife of a patient reported that she did try to keep herself active even if her husband could not always go with her, asserting, "I go out alone— sometimes I feel like I'm single. I try to go to the parties we're invited to because I feel that if we don't accept we're not going to be invited again." This spouse was attempting individually to maintain the social role responsibilities of a "couple."

Not all dialysis patients, however, reported restrictions in their social and interactional activities. It is interesting to note that those patients who did not live with family members or significant others reported themselves to be less restricted and more socially active. Overall, restrictions on social activities associated with the dialysis patient's condition were related primarily to two factors: (1) the machine—time restrictions and the necessity of a hemodialysis unit being available and accessible three times each week; and (2) patient fatigue or lack of energy—the patient simply feeling too tired or too "worn out" to be involved in social functions.

The Children

Relatively little has been written about the impact of a parent's condition of ESRD and hemodialysis upon the children. Salmons points out that generally "children of families in which one parent is chronically sick may become anxious or depressed and may exhibit behavior problems either at home or in school."[24] While no younger children were interviewed in the present research, selected parent–patient comments in the following discussion are reflective of the impact of the parent's perception of the parental illness upon their offspring.

A mother of two young daughters, while reporting their help and support, also pointed out the difficulties that they experienced. She reported that they became upset when she didn't feel well and was not able to do something for or with them, and commented, "It's hard on the children because I can't do as much as I'd like. Sometimes they say to me, 'Why can't you be like other mothers?' " This patient mother commented positively, however, that she felt her illness had made her children mature earlier. She observed, "They were very young and grew up with dialysis. My older girl has showed a lot more initiative than most her age. She takes a lot of responsibility for her younger sister." Another patient parent commented that she felt her children handled their anxiety by not talking about dialysis; and the spouse of a male patient described the impact of ESRD on her teenage daughter this way:

> She worries a lot about her father and sometimes I think she holds too much in. Sometimes he will be irritable with her when he comes home from work and she doesn't understand and then she'll get upset and run upstairs. Now that she understands it's his condition, it's a little easier.

The wife of a long-term patient reported that at times her husband got very angry with their daughter and seemingly over-reacted, especially if he was not feeling well. But she commented that she had always kept her daughter aware of her father's condition and added, "I've said to her, 'I know it's not always easy with daddy, but life is not easy, life is not a bed of roses.' " In general, study respondents did not report the perception of any more serious problems occurring among children due to one of the parents having experienced ESRD and long-term hemodialysis.

INVOLVEMENT WITH THE DIALYSIS TREATMENT
REGIMEN

Previous studies have suggested the importance of the influence of family and friends for successful maintenance of the overall hemodialysis regimen. Kaplan De-Nour and Czaczkes found that family attitudes greatly influenced the patient's behavior.[25] Cummings argues that, "Strong family support and backing is another essential factor in the adjustment of the dialysis patient."[26] He adds that family members, especially spouses, may require positive attitudes and attributes similar to those required by the patient, if the latter is going to adapt successfully to his sick role and treatment experience.[26]

Theodore Litman, in a study focusing on the family's role in physical rehabilitation, found a strong relationship between family support and therapeutic behavior. He noted that, "While 72.8 percent of those whose families reportedly offered encouragement and acceptance of their condition, received favorable ratings from the staff, the lack of such familial reinforcement characterized 76.7 percent of the patients whose performance fell below expectations."[27] In research whose purpose was to determine the relationship between medical condition, treatment compliance, and family interaction among long-term dialysis patients, Steidl et al discovered "significant correlations between ratings of overall family functioning and overall medical condition, and a near significant relationship between ratings of adherence to treatment and overall family functioning."[28]

Family member respondents in the present research reported a moderate degree of involvement with the patient's hemodialysis regimen, frequently admitting that they had been more concerned and/or more involved during the earlier period of the illness condition some 7 to 10 years earlier. The type of family involvement described usually centered upon (1) ensuring or fostering the patient's attendance at scheduled treatment sessions, and (2) supporting adherence to prescribed fluid and dietary restrictions. The taking of medications and care of access site seemed in most cases to be considered the responsibility of the patient.

In discussing the treatment sessions, the spouse of one 11-year patient expressed her concern about getting her husband to the dialysis unit during the winter (either she or her son drove): "I was so grateful—last winter wasn't too bad, but I always worry about the driving to the unit when it starts to snow." Another spouse expressed her difficulty in getting her husband to his treatment sessions when he became more physically disabled, as she did not have the strength to lift him in and out of the car. She noted, "You have to plan for these treatments. It's a constant worry." The mother of a female patient reported that sometimes her daughter just didn't "feel" like going in to the dialysis unit, explaining, "I try to encourage her to go. I tell her it's not so bad. 'Once you just get up and get dressed and go in, you'll feel better.' " And the grandson of an older patient discussed the fact that sometimes his grandmother just decided she didn't want to go for treatment. He said, "Then we have to give her a little push. We say, 'Come on, Mama, you got to go in there or the fluid will make you sick like you were before.' We just do a little fussing, and a little coaxing, and before long she goes and gets ready."

In regard to supporting the patient's adherence to the prescribed dietary and fluid restrictions, most family members reported continued involvement. Often, however, comments were made to the effect that both patients and family members had learned how the dietary restrictions could be modified to some degree and they were not as "strictly" adhered to as they had been early on in the adaptation process. A wife complained that her patient husband really had not been getting enough food during his first few months on dialysis until finally a "nurse" told her she could give him "a treat once in a while." She admitted, "Now sometimes if he wants a little ice cream, I give him a little, or some bacon—maybe I give him just one piece. That diet isn't worth eating if you just stick to it all the time." She continued, "When they first gave us the diet at the dialysis unit, for the first three months we really couldn't get going. Then there was this doctor, and he said to me, 'Does A. like bacon?' and I said, 'Yes.' Then he said, 'Go home and fry him up a pan of bacon. It won't kill him.' " This respondent issued the following warning to the interviewer, however: "You can't always tell new patients about this—about breaking the diet—they have to learn how to do it." The housemate of a female patient described her companion's behavior this way:

> She eats a lot of things she probably shouldn't, but we know just how far to go. She can still enjoy some of the things she used to. I know some things are definite no's, but why be totally denied? So once in a while we'll splurge and go on one of our benders. Sometimes I think maybe that's what's kept her going. We do eat extra kinds of stuff, but we have our restrictions, too. We don't go overboard even when she wants to. Sometimes I have to get on her and she won't eat it, but she'll rant and rave and then we'll start all over again. But people can't live on bean salad and jello all the time. They just can't. You have to have something really solid to hold onto once in a while, and we do.

A few patients, however, did report a rather more strict adherence to their prescribed dietary regimen and appeared to be supported in this by family members. The mother of a patient commented, "I cook just the things he's supposed to have because he hangs on to that diet—like an 'old maid,' he hangs on to it."

In regard to family members' wish for knowledge about and involvement with the dialysis patient's condition, a patient's wife summarized, "Right now I know as much as I want to know. I especially want to know if something happens, 'Why did it happen?' and then, 'What might come next?'

What I should expect next. How do I deal with this? These are the kinds of things I need to know."

RELATIONSHIPS WITH THE DIALYSIS UNIT STAFF

Overall, most family member respondents reported little interaction with the dialysis unit personnel, other than an occasional phone call. One patient's mother commented on what she perceived as the very positive benefit of discussing the patient's condition with "professionals" such as the hemodialysis unit caregivers and reported several conversations about her daughter's condition. A number of respondents, however, reported having virtually no contact with the unit staff and one spouse noted, "They don't listen to you when you try to tell them something. They'll shut you out because you know too much. You know certain things that they might not know and if they would listen to those things it might help the patient."

Dialysis unit caregivers validated the fact that minimal interaction occurred between staff and patients' family members. One head nurse observed that family members did call occasionally and that "usually a spouse or maybe one of the children feels comfortable in talking to a nurse." But she added that actual interaction generally occurred only on social occasions. She noted, "We had a party for the patients and they brought all their family members, so then everybody got to talk and know each other." Family members rarely reported spending much time in the dialysis unit itself, however. A head nurse gave this explanation: "Many family members really do not like to come in here. It's very difficult to watch your relative's blood sitting in a machine next to them."

Several family members interviewed described their social relationships with the families of other dialysis patients. Usually these relationships consisted of somewhat sporadic and infrequent interactions, however, rather than being characterized by more consistent friendship-type behaviors. The spouse of one 12-year veteran hemodialysis patient described it this way: "We keep in touch with some of the other patients and their families, but you know a lot of them who were first with C. at the unit have gone. Sometimes you're afraid to call. I tell them we're still here—we're hanging in there—we'll keep fighting."

BURNOUT

Much has been written about the stresses that are placed upon a family with a member being maintained on the hemodialysis regimen for ESRD. There have been few reports, however, related to the long-term (8–12 year) impact of such tension and/or disorganization on the family functioning. In the present research on long-term adaptation to maintenance hemodialysis, family member subjects discussed the results of the stress that they experienced over time and several respondents themselves employed the label "burnout" in describing their reaction.

Many family members suggest that the most difficult stressor that they encounter is the uncertainty of the patient's condition and prognosis. A mother commented that her daughter's illness condition was a constant worry and that whenever the phone rang she feared that it might involve the kind of rush trip to the emergency room that she had experienced several times. The spouse of a dialysis patient related the impact of her husband's condition as follows:

There have been times that things weren't going well; he'd get fluid in his chest and he couldn't even stretch out in the bed. It would really be quite upsetting to me. I couldn't even sleep. I'd worry that if I dropped off to sleep . . . I'd say, "Oh, Lord, what if he dies while I'm asleep." Sometimes I'd go into the guest bedroom and just cry, but he has never seen me cry because I feel that I have to be strong in order to keep him going.

Another wife commented,

At first when he went on dialysis I only half-slept at night. But it's still a constant worry. I go to bed with him on my mind and I wake up the same. It gets harder as you go on. They [the patients] kind of get tired. He'll say, "If I just didn't have to go in to that unit today, I could make it. I just don't feel like goin'." I'm the one who has had to really support him over the years. It's been hard. A lot of things you would love to do, but you just couldn't leave.

The daughter of a patient described the strain of powerlessness and her frustration at being unable to alleviate her mother's discomfort. She said, "It's very hard 'cause when she comes in [from the unit] she have to take her medicine and lay down and rest. I hate seeing her go through this every day. There's nothing I can do." The son of a long-term patient verbalized his concern about the possibility of his father's death this way:

I think my Mom would be able to cope with it. I'm sure she's thought about it. She's had to live with it for a long time. And in a different way than the rest of us because she is much closer and she does—she has lived with him longer than we have. It would really be tough on me, though, because I wouldn't have Dad there, because he has been an inspiration—although I've met other people who've gotten degrees out of their ears—but to me they're not as smart as my Dad.

He added, "It's interesting, I was watching *Knots Landing* last night and the same thing happened when one of the dads died and his son had to take over. And, you know, I could really identify with that."

Several family members perceived not only the stresses of constant worry about their dialysis patient, but also those related to constant care as being the catalyst for their "burnout." Interestingly, two female spouses commented, almost identically, that when their dialysis patient husbands were feeling poorly, they would complain loudly about their wives leaving the house for too long, even to do the routine grocery shopping. Both respondents commented that when they returned their husbands were angry and both also quoted their husbands as referencing the fact that the shopping trip shouldn't have taken so long.

One spouse of a patient who had suffered notable physical debilitation over time described her frustration:

It really is hard sometimes. I get so frazzled. I wonder really if I can make it. Sometimes you do feel just like walking out. One day he got so bad that I just walked out of the room and slammed the door and I said, "R., I'm not coming back until you can act differently."

She added, "I was really afraid I was getting burned out."

Another patient's spouse commented, "He [the patient] is very moody, and you have to understand this, but sometimes you just want to run away." She added, "My aunt said, 'How in the world can you stay here with a sick man,' but ain't nobody else to do it. It haven't been easy." A relative of a long-term patient reported, "I'm just tired, but there's no one else to do it [care for the patient]."

The mother of a patient described her experience with her daughter: "It's really been hard because she's had lots of ups and downs, and everything she goes through, I die with her so then she bounces back and she's living and I'm dying." She went on to express her fear that some day she might lose her daughter, and explained her attitude of detachment or pulling back from the relationship because of this. She put it this way:

You begin to get a burnout. Your feelings just aren't there, you begin to lose patience and all at the same time you still have your own problems. But as you get older, you can't live for other people. It's the same with M. I've come to that conclusion. I can't live for her. I learned from my two sisters; one was the older one and I was very attached to her, and I found that when you were too close to people, it really hurt when they passed; and I said I never want to get in a position like that, and this is what I think of as far as M. is concerned. I feel that she can take care of herself, there's nothing really that I can do and so I'd best get on with my life.

The respondent added, "You see, I decided that I'd better find a way to detach myself from this problem for my own sake." Finally, a dialysis unit head nurse associated the burnout phenomenon among families to the "long road" of ESRD. She observed, "Way down the end of the road in ESRD, frequently you see the significant others dropping off. Fade out. They fade out. They are tired of the chronic illness. They don't want to face up to it. Maybe they have been burned out in the real sick phase."

GUILT

Guilt as a stressor for dialysis patient family members has been focused upon to some degree in the extant literature. Brundage asserts that the dialysis patient's family may experience both depression and anxiety, and adds that "hostility toward the patient, if he is dependent, irritable, and childlike is not uncommon . . . hostility in the patient may cause the family to react with hostility and then feel guilty."[29] The concept of family guilt is supported also by Eccard, who notes that the "entire family structure may be in flux as a result of the illness."[30] Gutch and Stoner agree that the dialysis patient may react hostilely toward the family and suggest that, "hostility begets hostility, and family members become hostile toward the patient. They then feel guilty because of having these feelings."[31] Hostility-linked guilt among dialysis patient families has been noted as well by Shambaugh and Kanter,[32] who discuss self-help groups for ESRD patients and their families, and by Brinker and Lichtenstein, who report on spouses and their stress.[33]

Family member respondents in this longitudinal study did not directly express feelings of guilt related to the patient's condition, possibly because they had experienced, with the patient, a period of long-term adaptation. During this time one might expect that some feelings of anger and hostility had been diffused. Several respondents did, however, note their frustration when the patients became demanding or "unreasonable" in their requests for attention. The mother of one patient expressed the feeling that she would have felt guilty if she had encouraged her daughter to proceed with a transplant that did not succeed. She commented that she was grateful she had not been involved in the decision. She also noted that she was planning to stay clear of a decision about the initiating of CAPD, saying, "It's her life."

ROUTINIZATION OF THE DIALYSIS REGIMEN

Certain family members interviewed spoke to the issue of routinization of the dialysis regimen in context of the normalization of their family lives. One spouse of a long-term patient noted, "Our life has to go on despite dialysis." Another commented, "I do as much as I can to make our lifestyle as normal as possible and live from day to day. We just see what happens next." In discussing her daughter, this respondent observed, "She grew up that this is the way our family is. It's not easy, but we can't change it, so we accept it. We work around it." A spouse also noted, "We try to focus off dialysis. You could go crazy talking about that all day." The mother of a male dialysis patient reported of her son's dialysis regimen, "Over the years I've just gotten used to it. It's just like a daily routine." A patient's daughter said, "I organize my life around that dialysis schedule. I do for her and then I come back and do what I have to do at home." A housemate asserted that living with a maintenance hemodialysis patient did not hinder her own activities seriously, as she had "incorporated the patient's routine into her own. She observed,

> This is an everyday experience for me, it's something almost routine. I know what has to be done and when to do it. And it doesn't take anything away from my day—from my own lifestyle. I've incorporated all these things into my own life. It's not like I have a separate entity taking care of another person: it's all one.

Obviously, routinization of the patient's regimen and normalization of family life are to be anticipated in a population were families have survived a long-term illness adaptation process lasting up to 11 or 12 years.

THE FAMILY MEMBER AS POTENTIAL ORGAN DONOR

Although study patients' personal attitudes toward kidney transplantation varied notably (see Chapter 8), the possibility of transplant seemed to be always kept waiting in the wings as a "saving grace" should access mechanisms fail, or the machine become intolerable, and peritoneal dialysis be unacceptable. Renee Fox, in her paper entitled, "Am I My Cousin's Keeper," discusses a paradigm of gift-exchange and suggests that when this is related to the concept of organ transplantation, the "gift of life," one "immediately sees how interesting and complex, uplifting and distressing, the process of giving and receiving can be for the medical professionals, patients, and patients' families who are involved."[34] Several family member respondents in the study volunteered the information that they would be willing to donate a kidney if necessary, but the patient had refused such an offer. One male patient reported proudly that his teenage daughter had offered him her kidney "all on her own" but that at present he was still hoping for a cadaver transplant. A 19-year-old daughter, whose mother had also refused her offer of a kidney, shared her feelings on the subject: "Well, I sit down and talk to her. There's not too much I can do. I offered my kidney, but she don't want it. She said something might happen to me. I didn't feel relieved because I wanted her to have it, I hate to see her going through this every day."

Generally, family members appeared willing to donate a kidney should they prove an acceptable donor and should the patient be willing to accept their gift. One family member reported that his tissue typing was not acceptable in match to the potential recipient and expressed some dismay over the fact. A patient, who had rejected a cadaver transplant, commented, however, that his family had not "come through" with an offer of a donor kidney. He especially pointed out his sister's anxiety about the matter: "My sister doesn't get in touch with me but every once in a while anymore. She's afraid of getting involved with giving me a kidney."

It has been reported that close friends of dialysis patients occasionally volunteer for tissue-typing preparatory to possible transplantation. Although the kidney from a non-blood relative may not prove an acceptable match for surgery, the offer itself raises intriguing questions about the roles and responsibilities of these "kin-like" friends of the dialysis patient.

Several study patients identified as their most significant other a person such as a long-term housemate of the same sex or a neighbor of the opposite sex. These partners or "kin-like" friends were reported to provide care and support relative to ESRD and the treatment regimen within the context of a larger sharing of the patient's life activities. They, more than any other persons, most probably are aware of the patient's anxieties, concerns, and wishes relating to medical care and the uncertain future. Yet the formal or legal implications of such a role and responsibility in these kin-like relationships remains a gray area. Generally, a blood relative, spouse, or legal guardian is the only one permitted to sign medical care (or medical/surgical procedure) permits in health care facilities. Visiting privileges in certain intensive care settings may be restricted to "immediate family," the definition of which does not always include close friends. Important information about the patient's condition is often not communicated to such partners or friends, resulting in their increased anxiety and concern about the illness condition. Thus, dialysis patients' significant others in a nonlegally or nonsocially sanctioned relationship may experience unique stresses when dealing with the health care system and the provision of services for the patient partner.

Jane Howard, echoing Frost, observed that families provide a place where, "when you go there, they have to let you in, and where at the very least you can waken without surprise."[35] Families of dialysis patients, whether of the nuclear, extended, or friendship type, provide the above-referenced haven of security that may contribute significantly to the adaptation and, in fact, continued survival of the ill person. Although the stresses of coping with a long-term illness necessarily take a toll on the contributing family members, most accept the imposition willingly and attempt to maintain some semblance of normalcy in their daily routine. Supported by such attitudes and behaviors, patients are allowed the continued satisfactions and fulfillment of ongoing participation in the usual activities of communal daily living.

REFERENCES

1. Murdock GP: Social Structure. New York, Macmillan, 1949, pp 2–3
2. Horton PB, Hunt CL: Sociology. New York, McGraw-Hill, 1968, p 215
3. Goodman N, Marx GT: Society Today. New York, CRM/Random House, 1978, pp 339–342

4. O'Brien ME: Hemodialysis and Effective Social Environment: Some Social and ~~social Psychological Correlates of the Treatment~~ for Chronic Renal Failure. Unpublished doctoral dissertation, The Catholic University of America, Washington, D.C., 1976, p 6

5. Kossoris P: Family therapy as an adjunct to hemodialysis and transplantation. Am J Nurs 70:1730–1733, 1970, p 1730

6. Sand PS, Livingston G, Wright RG: Psychological assessment of candidates for a hemodialysis program. Ann Intern Med 64:611–621, 1966

7. King S: Social–psychological factors in illness, in Levine S, Reeder L, Freeman H (Eds): The Handbook of Medical Sociology. Englewood Cliffs, N.J., Prentice-Hall, Inc., 1972, pp 129–147, p 144

8. Litwak E, Szelenyi I: Primary group structures and their functions: Kin, neighbors and friends. Am Sociol Rev 34:465–481, 1969, p 469

9. Parsons T, Fox R: Illness, therapy and the modern urban family. J Soc Issues 8:31–44, 1954, p 34

10. Richardson HB: Patients Have Families. New York, Commonwealth Fund, 1945

11. Litman T: Physical rehabilitation: A social psychological approach, in Jaco EG (Ed): Patients, Physicians and Illness. New York, The Free Press, 1972, pp 186–203, p 200

12. Sussman MB: Family unit critique of selected scales and indexes available for measuring the relationship of family behavior to the etiology and course of chronic illness and disability. Unpublished draft, Project 94u44, Association for the Aid of Crippled Children, 1959, cited in Theodor Litman T.: Physicial rehabilitation: A social psychological approach, in Jaco EG (Ed): Patients, Physicians and Illness. New York, The Free Press, 1972, pp 186–203, p 200

13. Litman TJ: The influence of self-conception and life orientation factors in the rehabilitation of the orthopedically disabled. J Health Human Behav 6:249–257, 1965

14. Czaczkes JW, Kaplan De-Nour A: Chronic Hemodialysis as a Way of Life. New York, Brunner/Mazel, 1978, p 161

15. Duff R, Hollingshead A: Sickness and Society. New York, Harper and Row, 1968, p 251

16. Hampers C, Schupak E: Long-Term Hemodialysis. New York, Grune and Stratton, 1967, p 147

17. Viederman M: Adaptive and maladaptive regression in hemodialysis. Psychiatry 37:68–79, 1974, p 68

18. King S: Social–psychological factors in illness, in Levine S, Reeder L, Freeman H (Eds): The Handbook of Medical Sociology. Englewood Cliffs, N.J., Prentice-Hall, Inc., 1972, pp 129–147, p 145

19. Sorensen E: Group therapy in a community hospital dialysis unit. JAMA 221:899–901, 1972, p 900

20. Bailey G: Psychosocial aspects of hemodialysis, in Bailey G (Ed): Hemodialysis: Principles and Practice. New York, Academic Press, 1972, pp 430–440, p 434

21. Cummings J: Hemodialysis: Feelings, facts, fantasies—the pressures and how patients respond. Am J Nurs 70:70–76, 1970, p 71

22. Brundage DJ: Nursing Management of Renal Problems. St. Louis, CV Mosby, 1976, p 135

23. Brinker KJ, Lichtenstein VR: Value of a self-help group in the psychosocial adjustment of end-stage renal disease clients and their families. J AANNT 8:23-27, 1981, p 24

24. Salmons PH: Psychosocial aspects of chronic renal failure. Br J Hosp Med 23:617–621, 1980, p 620

25. Kaplan De-Nour A, Czaczkes JW: Personality factors in chronic hemodialysis patients causing non-compliance with medical regimen. Psychosom Med 34:333-334, 1972

26. Cummings J: Hemodialysis: The pressures and how patients respond. Am J Nurs, 70:70–76, 1970, p 75

27. Litman TJ: The family and physical rehabilitation. J Chronic Dis 19:211–217, 1966, p 215

28. Steidl JH, Finkelstein FO, Wexler JP, et al: Medical condition, adherence to treatment regimens and family functioning. Arch Gen Psychiatry 37:1025–1027, 1980, p 1025

29. Brundage DJ: Nursing Management of Renal Problems. St. Louis, CV Mosby, 1976, p 135

30. Eccard M: Psychosocial aspects of end-stage renal disease, in Lancaster LE (Ed): The Patient with End-Stage Renal Disease. New York, Wiley, 1970, pp 61–81, p 70

31. Gutch CF, Stoner MH: Review of Hemodialysis for Nurses and Dialysis Personnel. St. Louis, CV Mosby, 1975, p 187

32. Shambaugh PW, Kanter SS: Spouse under stress: Group meetings with spouses of patients on hemodialysis. Am J Psychiatry 125:100–108, 1969, p 102

33. Brinker KJ, Lichtenstein VR: Value of a self-help group in the psychosocial adjustment of end-stage renal disease clients and their families. J AANNT, 8:23–27, 1981, p 24

34. Fox RC: Am I my cousin's keeper? Annual Samuel Belkin Memorial Lecture, Albert Einstein College of Medicine of Yeshiva University, Bronx, New York, March 24, 1980, p 10

35. Howard J: Families. New York, Simon and Schuster, 1978, p 26

4

The Caregivers

The healer has to keep striving for . . . the space . . . in
which healer and patient can reach out to each other as
travelers sharing the same broken human condition.
                                        Henri Nouwen
                                        Reaching Out

Whence come those individuals who choose to spend their days caring for
end-stage renal disease patients who must face daily the stress and the pain
associated with their life-threatening illness and its treatment regimen? Why
do they choose to work with maintenance dialysis, knowing that, other than
kidney transplantation, there is no surgical procedure or miracle drug to
completely alleviate the patients' condition; knowing that as caregivers they
will be called upon to give of themselves and of their time in support of a
group of patients who must continue to struggle with their condition day
after day, year after year, merely to suvive; knowing that the patient-caregiver
relationships upon which they embark may well be cast in the context of
"'til death do us part"?

In the course of the longitudinal study upon which this discussion is
based, 45 professional and paraprofessional dialysis caregivers were interviewed in order to examine such issues as the role of the long-term dialysis
caregiver, patterns of staff–patient interaction, caregivers' attitudes toward
patient death, staff members' stresses and frustrations, and the chronic patient
and caregiver's relationship.

The dialysis unit caregiver study respondents consisted primarily of
head nurses, staff nurses, and therapists. Data were also elicited from interviews with several physicians, social workers, machine technicians, and clerks
associated with chronic dialysis units.

CHARACTERISTICS OF THE DIALYSIS CAREGIVER

Staff member study respondents consisted primarily of baccalaureate and diploma-prepared registered nurses (several master's-prepared nurses were also included), licensed practical nurses, and paraprofessionals employed as therapists and machine technicians. Ages ranged from early 20s to late 50s, with the modal age ranging around 30 years. Although the nurses were predominantly female, many males were employed as therapists. Approximately two-thirds of the group were married and most had been working with dialysis patients for 5 years or longer.

The majority of caregivers interviewed noted that they had received special training in hemodialysis technology (courses of study varied) for approximately 6 weeks to 2 months before assuming full patient responsibility during treatment. Some worked 3-day weeks in shifts of approximately 12–14 hours, while others, particularly those in hospital-based facilities, functioned within the more traditional 40-hour week. Many caregivers, however, reported working "overtime" and some had "on-call" hours.

In order to determine what stimulates one to work in the arena of end-stage renal failure and dialysis, hemodialysis unit staff members themselves were asked to characterize a dialysis caregiver. The following examples demonstrate some staff members' self-perceptions: "I think you have to have intelligence. You have to have some skill technologically to be able to learn the treatment procedure, but especially you have to have a lot of patience because you're not going to see any kind of quick results"; "I think you also really have to be able to deal with death, too; there are some nurses who just can't handle it and they eventually leave"; "I think one [the dialysis caregiver] has to have a realistic expectation of what the disease brings to the situation, in terms of the fact that it is a chronic disease; you have to be able to set goals that are appropriate for yourself in working with patients in terms of what the patient is going to be able to accomplish in light of what is ahead of him; you have to be flexible." A dialysis nurse has "to be compassionate." A dialysis caregiver has "to have empathy and skill in hemodialysis to be able to deal with death and dying—the emotional."

One dialysis unit head nurse couched her response in terms of what she called her "famous speech for people [prospective staff] I have to interview." She asserted,

I think it [working with dialysis] takes a person who can understand theory. I think it takes a person who is manually dexterous —who can manage things like ultrafiltration and learn the machines. I think is takes a genuinely empathetic person, who can tolerate long-term, ongoing relationships with patients. And the problem with that, as I see it, is that you have to maintain some kind of professional distance from the patients. But you see the same people over and over, so many times, that it is difficult to maintain that fine line between involvement and professionalism. So, it kind of tends to illness into a social relationship and when you come to a difficult point in that person's health or illness, you have to set limits and in that respect it is difficult. I think it takes an extremely introspective person especially [to work] in this unit.

A hemodialysis unit social worker described her perception of the dialysis caregiver's needed attributes this way:

I have been here a long time and I feel like I've seen a lot of nurses come and go. I have seen the ones who do well with the patients. You have to be extremely matter-of-fact about what is going on, because I know the people who are working directly with the patients, they go out of here exhausted. The ones who get upset and agonize over it seem to get overloaded and leave. The people who seem to stay the longest are those who can treat it [dialysis]—not coldly—somewhat objectively, be interested in the patients but don't let it be overwhelming to them.

A physician summarized his self-perceived requisite characteristics in these words:

I think a lot of patience [is needed], and hard work also. It takes a lot of time, you know, a lot of your time, not only from yourself but from your family. Getting called in the middle of the night; going to the hospital in the middle of the night. And really that is what it is all about, a lot of patience, hard work and understanding of the patient's human needs.

While the above comments reflect perceptions of the hemodialysis caregivers' need for intelligence, technical skill, and a realistic perception of ESRD, the more predominant response themes relate to such interpersonal and caring characteristics as patience, understanding, compassion, empathy, and a commitment to long-term relationships. Many caregiver respondents also highlighted the need for caregivers to be involved yet objective in order to avoid burnout and maintain their own emotional stability.

THE ROLE OF THE PHYSICIAN

Ultimately it is the physician who determines a patient's need for dialysis and "prescribes and conditions by which it will be conducted."[1] In general, direct care-givers in the dialysis unit who were interviewed in the study reported positive relationships with their unit physicians. Most seemed to view the physician's role as not central to the carrying out of the treatment procedure itself, but as related to direction and management of the patient's total dialysis regimen. One hemodialysis unit head nurse suggested that there occasionally was some tension in the unit "if the 'docs' are acting up." She added, "Sometimes I think they [the physicians] don't understand the strees we're under or the responsibilities we carry." Another unit head nurse noted, however, that in her unit the nurses, therapists, and physicians all worked very well together as a team:

In this unit our physicians are all part of a multidisciplinary group. We have a unique situation here. I have worked in other units where the physician wasn't there much . . . but here it is different. Nurses and technicians have a great deal of input and the physicians do, too. They all talk about the patients and work with the patients closely.

A physician described his caregiving role this way:

I get emotionally involved with their [dialysis patients'] problems, especially those that become kind of close to you—they confide in you with their problems. Young patients I get involved with, too, sometimes. But I try not to, because this can make it very difficult to treat them.

In discussing emotional involvement and the nurse–physician relationship, he added,

The nurses [can] get emotionally involved with the patient to the point that they are always on your back—"why don't you do this, why don't you do that?" It may be a healthy way of doing it. But if it is done to the extremes, it is not good for the patient. Because you will be using your heart, not your head for the management of the patient.

This physician added that it was sometimes difficult working with the dialysis population because of the chronic nature of the illness and the rapid deterioration of certain patients. It might be pointed out that over longer periods of time physicians involved with dialysis cannot avoid experiencing the stress and the pain related to their patients' uncertain futures and often unstable course of illness and treatment.

THE ROLE OF THE NURSE

Gutch and Stoner suggest that the nurse has played a major role in maintenance hemodialysis since its inception in the 1960s. They add, however, that the role depends in part upon "the locality, size and nature of the facility—federal, university, municipal or private."[1] Nurses may be employed in dialysis units as coordinators or supervisors, with administrative responsibilities related to scheduling and planning of care; as educators, involved in both patient and staff teaching; and as team members, providing direct patient care during the hemodialysis procedure.

Standards of nursing practice were originally adopted by the American Association of Nephrology Nurses and Technicians (AANNT) in 1972; and "these have since been accepted by the American Society for Extra-Corporeal Technology and are recognized by the American Medical Association,

American Society for Artificial Organs, American Nurses Association and other organizations basic to nephrology nursing."[1]

Most nurse respondents in the study, despite their specifically identified roles and role activities—e.g., head nurses, shift supervisor, team member— reportedly perceived their activities as being ultimately related to promoting and facilitating patient adaptation to maintenance dialysis and the total treatment regimen. Many articulated role responsibilities related to patient and family education, as well. Following are selected nurses' comments relating to their personal perception of caregiving responsibilities: "I think that probably [in dialysis nursing] the least amount of effort involved is technical. Once you learn the machine, it becomes very routine. When you first go in you're wrapped up in that machine and you know, it's really exciting. That passes very quickly. And then you get into emotional and social interactions with the patients"; "I think psychological support and helping the patient and family adapt [to dailysis] are the important part of the nurse's role; I think the technical skills are probably not all that necessary because the procedure itself is so routine that once you learn to do it, even though at first it might take you a little bit longer, you will never forget it. You can do it in your sleep."

One unit head nurse commented that while she viewed the nurse's role in the procedure as primarily technological, more time could be spend on psychological support for the patient—"if you had the time to do it." She added, "Many times you don't, you know." Another respondent described a role distinction between the nurse caring for the chronic versus the acute dialysis patient:

It's a combination [of physiological, technical, psychological, sociological support] I think, depending upon whether the patient is an acute dialysis patient or a chronic dialysis patient. With acute patients there may be more crises or problems on the machine. But we're finding with our chronic patients that we are having to be more geared to their psychological needs which, in part, may be due to the fact that they are so stable on dialysis and we're less involved with the technical aspects of the treatment.

One nurse who had been working with dialysis and transplant patients for over six years, focused upon the nurse's direct caregiving role and responsibilities in distinction to those of the physician. She stated,

Kidney disease is a really difficult condition to deal with. Patients are so dependent on the caregivers and on their caregivers' moods, however they feel. Their caregiver is their means to life. The nurse is really sustaining their [the patients'] lives. While the patient is on dialysis the nurse could cause that patient's death. So, you could differentiate the nurse's role as caregiver in that respect from the physician [role]. Patients might see the physician once or twice a week, but the nurse is there all the time. The nurse is responsible for everything, for that patient's life and for that patient's being able to go home. When you look at how much [care] is the nurse's responsibility, it's sort of frightening.

THE ROLE OF THE THERAPIST/TECHNICIAN

The dialysis therapist or technician is a skilled paraprofessional caregiver who has been trained in either the hemodialysis treatment procedure or the technical maintenance of the machine (artificial kidney) or both. Some dialysis units have separate roles for the "therapist"—direct care-giver— and for the "machine technician"—person responsible for the care and maintenance of the technical equipment necessary for the treatment procedure. Training for paraprofessional caregivers varies from unit to unit, but generally involves classroom teaching and a supervised clinical practicum of up to several months' duration.

Therapists interviewed in the present research generally perceived their role as related to direct patient care from both a technological and a psychosocial perspective, as some of the following comments illustrate: "Initially, I think we have to focus on the technical aspect of the machine, but after you get comfortable with that you can begin to get more involved with the patients—working with their specific problems and listening to them,"; "After a while you just get so used to the machine you can do it automatically and then you can be more emotionally involved with the patients and supportive of their needs and problems. That's an important part of our role."

One therapist highlighted her perceptions of the import of continued care and accuracy in regard to the physiological/technical aspect of hemodialysis, however:

You get to a point where you are used to doing it [the dialysis treatment] and you can deal with the situation and it goes faster, and you tend to be more accurate. You get more accustomed to it. But I don't think you ever totally relax in the situation because you've got the patients there. You've got to monitor them. You've got to watch them all the time. You've got to watch the machines. You've got to know the equipment. You've got to know the patients. We have patients, that when they yawn, I panic. I mean, it's an immediate response. You jump on them because a yawn is an indication that if you wait two seconds, they're going to be out [unconscious]. And then there are other patients who can have a nothing blood pressure, but be perfectly fine.

Overall, the caregivers in the study, physician, nurses, and therapists, perceived their roles as having dual foci: physical/technical and psychosocial. While the primacy of either focus varied among respondents, a holistic approach to dialysis patient life adaptation was the norm in terms of caregiving attitudes and behaviors.

THE ROLE OF THE SOCIAL WORKER

As with the nurse, the social worker's role may assume a different character from unit to unit depending upon such factors as patient case load, hours worked (whether full-time or part-time), expectations of the physicians, administrators, and caregiving staff, and the individual social worker's own preparation and experience in the field. Generally, social workers reported that they tried to spend a period of time at consistent intervals with each of the patients in their assigned dialysis unit or units; most admitted, however, that heavy caseloads often precluded such routine patient visits, and functioning often tended to be primarily crisis- or problem-oriented.

The social worker may function alone or as a member of a social or social-psychological or psychiatric support team. One psychiatric social worker described her role this way:

We have two social workers on board here. I work with the psychiatrist and provide [patient] followup. The medical social worker is responsible for all of the intake, all of the new patients that come in. She initially screens these patients for purposes of selection into the program and provides them with any type of resource they might need, or referrals for resources. Placement is part of her function. I am primarily responsible for dealing with psychiatric problems with the patients and the medical social worker carries out the concrete tasks in terms of referral, educating the patient about the system and those kinds of things.

In discussing her relationship with the caregiving team, the social worker added,

The nurses here are coordinators of the care planning for the patient and all the other professionals play a role in that care planning, but we are not in charge of it. We participate in it. But we have nurses and we also have technicians. Oftentimes, I find that the nurses are pretty well able to independently set up a plan, whereas with the technicians, I might have to take a larger role in helping to implement the plan, especially if there are a lot of psychological issues that have importance in terms of the patient.

Another social worker who had been involved with dialysis patients for over five years, admitted that, after a long period of time in the field, frustration, boredom, and a kind of burnout can occur. When asked how this was handled, she responded, "Well, you sort of withdraw, I think. You are less active at certain times than you are at others. For the sake of your mental health you kind of pull back and focus on something that is not too stressful and talk to some patients who you know are not going to present you with horrible problems." In response to a question about the most frustrating aspect of her role, the respondent added,

Since I have been here for five years, I feel like many times I am saying the same things to the same patients, month after month, and answering the same questions, and trying to help them sort out how to solve the same problems. It is like it just repeats and repeats and repeats. Somehow, they haven't really learned or I haven't been a good teacher or the same problems just recur and recur and recur, whatever it is. It seems like problems happen over and over and over again and there is never an end to it. That's what burns you out!

Another social worker pointed out the need for periodically insulating oneself from the stress of day-to-day contact with the hemodialysis patient group and their problems:

I've been a dialysis caregiver for over three years. That's a long period of full-time. We've gone through a lot of changes in our units and at one point I found myself just going through the paces of talking to patients and sitting in my office; and my desk was getting messier and messier. I'd say, "I've got to clean this off," you know. "I've got to be all things to all people." All those needs out there—there's nobody really to fill them. And out of my frustration, I handled it by just withdrawing for a while, not even doing what I should have been doing. You just get through the day and begin to think, is this your form of burnout? And then the most amazing things happen, and you get excited again and into what you love and you're okay because you've had a respite.

In discussing the stresses of long-term work with maintenance dialysis patients, a considerable number of staff members either directly or indirectly, touched upon the paradox of their own needs, as related to the needs of those for whom they care. Many verbalized a periodic longing to "get away" and forget about the suffering and pain associated with their work. Yet, dialysis caregivers were found to be consistent and committed to the care of their patients, for the long run. In assessing the experiences of the care-givers, one is reminded of Henri Nouwen's discussion of the "Wounded Healer," the one who must look after his own wounds, but at the same time be prepared to heal the wounds of others.[2] Nouwen's concept, which sug-gests that one must anticipate the time when he will be needed by others, is abstracted from a Talmudic description of the Messiah, who is described as follows:

He is sitting among the poor covered with wounds. The others unbind all their wounds at the same time and bind them up again. But he unbinds one at a time and binds it up again, saying to himself: "Perhaps I shall be needed: if so I must always be ready so as not to delay for a moment."[3]

CARE SETTINGS

The in-center hemodialysis treatment procedure may be carried out in a variety of settings. Dialysis units may be small and intimate facilities with as few as five or six patients receiving treatment at any given time or they may be large centers where as many as 40 or 50 artificial kidneys are available for each treatment shift. The procedure can take place in an in-hospital acute care or acute/chronic dialysis unit or in a free-standing commercial outpatient center. Although, as previously noted, caregivers did articulate some differences in regard to stresses in the different units, both for patients and staff, there are many commonalities. One dialysis unit head nurse noted the difficulty of having to provide care in a large, open treatment room where the caregiver was always "on center stage" in full view of the entire group of patients. She commented, "Boy, patients do watch everything you do, too. You can't escape or hide. You're just out there." A therapist added, "You have to be constantly on guard watching those lines—you can't goof off when you're in a section because it's just then that somebody would have a problem and you might not see it." The care setting for the hemo-dialysis treatment procedure is somewhat analogous to that of intensive care, where crises may occur with little or no warning and constant vigilance by caregivers is requisite to patient survival.

One setting risk frequently mentioned by the caregivers was a possible health hazard in regard to contracting hepatitis. Procedures regarding hand-washing and the use of utensils and food are carefully spelled out and en-forced on most units; however, as one nurse, herself a recent hepatitis patient, ruefully noted, "We do take precautions, but still almost every nurse around here has had it." Most caregivers did not appear overly concerned about the possibility of contracting hepatitis, but considered it, in the words of one therapist, "just an occupational hazard."

Caregiver respondents reported that a change of care setting, e.g., transfer from a larger cosmopolitan dialysis unit to a smaller provincial fa-cility, occasionally proved productive for alleviating stress or the symptoms of impending burnout. While staff personnel tended to stay in the field of hemodialysis care over time, a fair amount of movement between units was observed during the course of the study.

STAFF PREPARATION, EXPERIENCE, AND TURNOVER

Caregiving staff working in a dialysis unit may include master's, bac-calaureate, associate degree, and diploma-prepared registered nurses, as well as licensed practical nurses and skilled paraprofessional therapists and tech-nicians. In general, the nurses and therapists interviewed noted that they had received adequate training in regard to the technical aspects of their work, but many verbalized a felt need for some type of in-service education or staff development program to help them better understand and cope with their patients' many social and psychological problems and needs. As one master's-prepared registered nurse put it, "We do get to go to conferences [on ESRD and related problems] sometimes, maybe once or twice a year, but we need something to really stimulate us to do more for the patients. Sometimes we just get in a rut and kind of bored with it all and we could be spending more time helping the patients cope with their problems." Some caregivers, however, did report the regular occurrence of in-service education to review patient problems—with discussions involving such team members as the physician, psychiatrist, social worker, and dietician. The positive benefit of these group explorations was verbalized.

It was not difficult, for the purposes of the study, to find dialysis caregivers who had been working in the area for as long as five years. As one nurse stated, "You either love renal nursing or you hate it. When you start you either turn off and quit right away or you just stay on forever. It gets in your blood. You have to really like these patients, but once you do you're hooked." Despite this and similar statements by caregivers interviewed, it was noted that many dialysis units have a high rate of staff turnover—among both nurses and therapists. When turnover occurred, however, the caregiver often did not entirely leave the area of dialysis or renal nursing but simply moved, as noted previously, from one unit to another or from one type of dialysis to another (e.g., acute to chronic hemodialysis). Thus, caregivers frequently brought a number of years of experience and much expertise to a new setting.

Those who felt the situation of staff turnover most seriously, though, were the receivers of care, the dialysis patients. Many patients expressed their concern about staff changes in their own unit and verbalized the anxiety that this situation caused for them. As one patient put it, "These nurses and therapists—they're our security. It's real hard when you have to constantly get used to new people. You just build a relationship and somebody's gone. And it's really hard when the head nurse leaves. That gets all the patients very nervous."

"ELITISM" OF STAFF

"Elites" are defined as "influential, expert or powerful minorities,"[4] and as "a privileged group exercising the major share of authority or control within a larger organization."[5] Proceeding from such a sociological perspective, hemodialysis unit staff members may be considered "elites" in that they are a small group frequently viewed as special or expert within the larger caregiving system. They may also be said to be particularly elite when functioning within a hospital system, as they generally exercise almost total control over the hemodialysis procedure and within the hemodialysis unit—which sometimes does not even fall under the authority of the nursing supervisor. To employ the concept of elitism as understood by the nonso-ciological community, dialysis caregivers are considered by some, and, more significantly consider themselves, as unique or special, different from other caregivers in their knowledge of the treatment procedure, privileged in the responsibilities that they hold, and united as a group by those common distinctions. Such self-perception of elitism may be an important factor associated with the caregivers' job satisfaction, self-protection, and mutual support. In the words of one dialysis unit staff nurse,

> We're unique. There's a unity among us. No one can understand it. Even some people in the medical field don't understand it because they don't know what dialysis is. That's why I think that dialysis caregivers are a special group. We've often said the diversity among us socially, and financially and culturally, is so large here in the unit, but we still seem to click in most cases. I think that we're all a little bit different from the average medical person on the outside.

One head nurse of an in-hospital dialysis unit described her unit's position this way:

> As far as the rest of the hospital, I think they see it as "us and them," or "them and us." We are new—special—foreign to a lot of people. There is an attitude of "Who are these people?" People think we are treated special. And there is a sort of jealousy there. And also I am under administration. I answer to nursing, too, but not the same as the medical-surgical wards. If I make a decision or do something, I go to my boss with everything in hand. Whereas, with nursing they go through a much different procedure to do things.

Another head nurse described her unit's situation somewhat similarly:

> There's a lot of responsibility and there aren't many resource people in the institution that can help out with that. We're very different. The supervisors will come up, but they'll say, "Well, you know; I don't know. I don't know what to do." Or, "I'll come up but I won't know what to do." I think, too, there's a resentment from some of the other staff. We don't work Sundays, we don't work nights. But then they're not on call the long number of hours that we're on. They don't have to cover their staffing the way we do. But we do leave early in the evening when our patients are done.

Finally, a dialysis staff nurse pointed out the benefits of being a member of a special group. She observed, "We are a different, special kind of caregiver and we stick together and support each other. Other nurses don't understand the kinds of problems we have to cope with, but it makes you proud to know you can cope and do the kinds of things you have to do." Staff respondents both within and between dialysis units were very supportive and protective of each other. They often interacted socially, outside of the work setting, and shared problems and concerns in regard to family and other nondialysis-related activities.

STAFF SUPPORT SYSTEMS

It has been argued that formal "peer support and peer communication programs are an effective model . . . in the sharing and enhancing of positive or more growth-producing subjective impressions"[6] among dialysis/transplant unit nurses. Weekly nursing staff meetings both with and without the assistance of a consulting psychiatrist have been found useful as a continued means of reducing stress among hemodialysis unit staff members.[7]

Hemodialysis caregivers in the study verbalized quite forcefully their perceptions of the importance of both formal and informal support systems relative to their professional interactions and activities. Their preferences, however, for the type of support needed were varied. One staff nurse stated that originally she had liked the idea of a formal group sharing, but now preferred just to "ventilate with her friends," informally. She said,

A couple of years ago they [the unit administration] tried having some meetings. The social workers got together with us and we would all meet and talk about our concerns, but it didn't seem to work out too well. It just got very hostile, you know. It turned into a complaint session, not any kind of helpful support system and everybody would complain about this doctor or that doctor. It just disintegrated. And I guess no one felt like we were getting any support.

Several other caregivers reported similar problems with formal support groups and generally suggested that such "programmed sharing" didn't always work because the support was "scheduled" and not there if you were really in a crisis and needed someone right then. Other dialysis unit staff members, however, felt positive about the formal support systems provided to their facility. One caregiver reported that she was involved in regular bimonthly conferences to discuss staff and patient problems and these were found beneficial to all caregivers on the unit. In some cases both formal and informal kinds of support were reportedly provided to a dialysis unit staff. Such a system was described this way: "We have a clinical psychologist here and she comes up and talks to us about any problems that we are having or any problems that the patients are giving us. We also have Rev. A. and we can talk to him about any problems we have." Despite the support systems in place, however, most caregivers reported a need for continued understanding and concern on the part of their administrators and supervisors.

FORMAL STAFF–PATIENT INTERACTION

In analyzing formal or treatment-related hemodialysis staff–patient interaction, some concepts that emerged as predominant included control, bargaining, testing, manipulation, and dependence. It has been noted by Lira and Mlott that "the psychopathological effects of chronic hemodialysis treatment programs are widely recognized and frequently attributed to the loss of autonomy or control experienced by the patient."[8] Kroemeke and Nassar, in contrast, found that de-emphasis of staff control was perceived by patients as well as by staff members. They suggested that as patients had specified duties and responsibilities during the treatment, "therefore the staff have little need to exercise control in order to make the patients perform their expected duties."[9]

Although one often finds mini-conflicts or struggles for control between staff and patients within the dialysis unit, the present research revealed that from the patients' perspective, control or power over the dialysis treatment regimen resided primarily with the caregivers. One dialysis patient reported, "You have to do what the staff tell you because your life really depends on them." Relating an incident of nonwillingness to accept the physician's advice, she added, "The doctor wouldn't accept it [my decision]; so he just turned me off. He got mad. You have to do what they want." Other patient comments relating to control represented similar themes, "The machine is their responsibility. It's helpful if you get along well with them because you want monitoring. You have to have trust. They are the ones doing it." "Sometimes a nurse may not be in the right mood when she put you on and she talks to the patients in a bad way and every patient feels it. We try to come in here in a good mood and they can put us in a bad one." "Most of all patients feel downhearted about not having a doctor around all the time in case you get sick. You just have to wait. He's across town somewhere swimming or playing golf." Finally, one long-term dialysis patient commented, "You never can quite reach the professional staff. They have their own authoritarian attitudes."

Perception of staff control was also evidenced in the responses of some hemodialysis patients' family members. A spouse of one male dialysis patient complained, "He once had a really big problem with one of the therapists. He came home very depressed. I called and asked if they could switch his days but there's not really much you can do." A patient's daughter commented, "The patients biggest gripe is that the staff is in control and does not listen to them. Nobody can understand what they are going through, so how dare you tell them this or that. Dialysis patients know better than anybody in the world what's wrong with their own body—when their chemistries are off. But the doctors and nurses don't listen." Finally, a caregiver helpfully summarized, "I think there is a conflict over allowing the patient to have control. The anxiety, I think, belongs a little bit to the staff and a little bit to the patients. [It's related to] the fear that something might happen, and who's going to be responsible."

Although the above comments indicate a strong perception of staff control by dialysis patients, the care-givers themselves frequently articulated a perception of patient control or struggle for control in terms of the concepts of bargaining, testing, and manipulation.

Bargaining. Bargaining by dialysis patients was often described by caregivers and was exemplified by such reported patient comments as, "If you'll let me off the machine early today, I promise to stay extra all next week"; "I promise that if I can have one more cup of ice, I won't ask for anything else at all this morning." One caregiver admitted that she tried to work with patients' bargaining, explaining,

> If a patient says, "I want to come off [the machine] at 11 o'clock," I say, "Fine, but don't ask me again for a month." You can't just always go and do what the patients want, but sometimes if I find that they're not being unreasonable, I take them off.

Testing. Testing by dialysis patients was identified by the caregivers as relating to the patient's need for attention and caring. It was described by one therapist this way:

> Patients come in here maybe four or five kilos over, and they wait to see our reaction. They think if they come in grossly noncompliant or in fluid overload, that they're getting to the caregivers or that we'll get upset. They'll sit there and say to another patient, "I ate a banana last night," which is really cool except that if another patient tries it and it doesn't work, they are going to feel the fact that they could have been the cause of his death. I get so angry when I hear that and so in a sense they are getting at me, since I do get upset. Because I care. And they know that and they're testing me. Maybe they have to keep seeing if you do care. You know they push you to see how far you can go. And they like it when you yell at them because it means you care.

A hemodialysis unit head nurse summarized, "Sometimes they [the patients] test us; they are just trying to get our attention—to know that we really care about them."

Manipulation. Attempts at staff manipulation were reported by caregivers to be related to patient attempts to maintain some control and autonomy over their lives and their illness conditions. The following caregiver's examples illustrate the concept: "You know patient X. Well, sometimes he comes in [to the unit] at the time for his treatment and just walks around and talks to the staff and the patients and then just leaves [without hemodialysis treatment]. I get so mad. He thinks then that he can just come back in here whenever he needs us and he probably can. If he comes in the next day overloaded, he knows we'll probably run him"; "Sometimes they'll [the patients] not show up for a shift. So I yell at them. And run them high the next time. It gives them control, if they can skip a treatment once a week."

In line with the idea of patient's needs for staff concern and attention, a caregiver speculated, "I think sorts of attention-getting things are a way of manipulation for the patients. If somebody else is sick on dialysis, you need to give them a little extra attention so they [other patients] kind of become sick, too." Another staff member gave this example: "The patients all have their little ways. If they know, say, J. won't do what they ask, they'll call somebody else over as soon as J. goes in the kitchen or whatever, especially somebody that they know always says 'Yes' and never says 'No.' And they know! They know who is very stern and will make them do and they know who's lenient and will let them do what they want, give them some extra attention."

A head nurse described how her unit was attempting to handle the problem of patient manipulation:

> We really stress that everyone [on the staff] follow whatever objectives we have set so that everyone is treating the patients the same, and the patients don't perceive that one person can be easily manipulated while another can't. And that has worked to some extent. It usually takes time, however, for people to resist being manipulated. It's hard to be tough when someone is seriously ill. You can't tell whether they're manipulating you or actually speaking of a real physical problem, particularly people who decide they either want to show up for dialysis late or don't want to stay for the entire treatment. And it's hard to make that judgment.

Dependency.   Dependency to some degree is to be expected in the case of most chronic illnesses; it may be more or less destructive to the patient's functioning depending upon the severity of the individual's dependency needs. Patient dependency may be focused upon professional caregivers or upon other significant persons such as members of family-kinship or friendship groups. The present discussion focusses upon the patient's dependency on members of the hemodialysis care-giving staff. (Dependency upon significant others with whom one has primarily an affective relationship is discussed in Chapters 2, 3.)

Hemodialysis unit staff members' reported that patients' dependency needs and behaviors can be markedly different. For some patients, dependency was significant, as the following caregiver's statement demonstrates: "There are a certain number of patients who have a dependent kind of personality. They lean on caregivers and may be manipulative and take advantage of being ill, for a secondary gain." Then she added, "But I don't think that is true of all patients." Certain patients were described as "not wanting to do for themselves" in terms of self-care and the treatment procedure, "not taking responsibility for their dietary and fluid management," calling the dialysis unit frequently with concerns about their condition, and consistently requesting a particular nurse or therapist to initiate the treatment. A dialysis unit head nurse expressed the opinion that patient dependency on the staff was initially supported by the staff itself: "I think sometimes we kind of encourage that because we will have a new patient who is really over-whelmed with just so much new information and I think we try to, maybe, smother him a little at first." Another nurse elaborated: "I think dependency is partly our fault. We take their coats off. We hang them up for them, do everything for them. I think it's partly our fault because if we started the right way we wouldn't have these problems. It's too late now to start over again." A social worker, discussing the dependency issue, focussed on the staff members' responsibility and needs as well:

I think a lot of dialysis patients get overly dependent. I feel that a lot of patients become too dependent upon the staff but that, maybe, the staff need that as well. Sometimes it's easier to do something for someone than it is to try and train or teach the person how to do it for themselves. Especially in the chronic dialysis unit where people aren't really expected to do that much for themselves.

Several caregivers briefly discussed the patients' dependency upon family and friends; one therapist indicated, however, her belief that while long-term patients did become dependent upon their families, "90 percent of the patients become more dependent on the unit care-giving staff than on their own families." She elaborated: "You see the families sort of fading out as time goes by and the staff becoming more and more important." This statement is supported by findings in an early phase of this study, which examined quantitatively the relationship between social support and ad-aptation to maintenance hemodialysis over time.[10]

THE CHRONIC PATIENT RELATIONSHIP: STAFF FRUSTRATIONS AND BURNOUT

Most caregivers are aware of the general "'til death do us part" nature of working with chronically ill patients, and they are prepared for the "for better or worse" aspects of these relationships. Sometimes, however, the "for worse" characteristics of the patients or their illness condition seem to overwhelm the interaction. As one hemodialysis unit staff member put it, "Let's face it—dialysis is depressing, for them and for us, but this is going to be a long-term relationship and we have to deal with it." As suggested earlier, patience was frequently the most important characteristic for a di-alysis caregiver, verbalized by the staff members themselves. A staff nurse put it this way:

There aren't many rewards in renal disease at all. You see the patients go through so many changes. They lose so much of themselves, giving it up to their illness.

They go through all of your psychiatric defense mechanisms, so you have to understand those pretty well to have the patience to work with these patients.

Staff frustrations related to the hemodialysis patient care-giving situation most frequently revolved around the concepts of minimal or no physical improvement of patients, gross noncompliance with the therapeutic regimen, the repetitive or chronic nature of the work, and patient deaths. The following examples are illustrative of caregivers' feelings: "I think one of the greatest frustrations is not seeing any kind of results from working with patients for a long time. For the chronic ones it's very difficult because a lot of them have been on dialysis for so long. It seems like the ones you have the most gratification from are the people that are just starting dialysis"; "I think probably the most frustrating thing [about dialysis nursing] is that we basically don't see patients when they get better. If patients are transplanted and the transplant is successful, we don't see them again unless they just happen to stop by and visit. And if someone is in acute renal failure and recovers, then we usually don't do any type of followup with them. So what we see are the patients that either stay on chronic dialysis, or the acute patients that don't get better. We're faced with chronic illness all the time." A head nurse commented,

I have to tell you I think that this is a fairly depressing place to work. I think that renal disease is depressing. I think that you have to be psychologically at your best to function in this unit. At your best. I feel very strongly about that. And I think it is just difficult and depressing to see these people deteriorate, right in front of your eyes, when you know that the doctors are doing everything that is medically possible for them; it has a very depressing element to it.

In discussing the problem of gross noncompliance with the dialysis treatment regimen, most caregivers vented a real anger and frustration about what they considered suicidal or near-suicidal behavior. One nurse stated, "I think one of the most frustrating things is to see a patient who can't accept the program. Sometimes they act like they're accepting it but you know they're not. They come in here 4 to 5 kilos over and you know they're heading for disaster! I find this terribly frustrating." Another caregiver related patient noncompliance to lack of what she felt was needed professional support: "It is so frustrating to see these people continually be noncompliant simply because we are not together enough to provide them the structured kind of program and consistent support that they need. And that makes me feel inadequate, not personally, but just system-wise."

Many caregivers mentioned the frustration of the repetitiveness of their work in terms of patient psychosocial support activities as well as the technical aspects of the therapeutic procedure. One staff nurse noted that the "day-in and day-out coping with the same patients and the same problems can really wear you down." She added, "It's easy to get burned out in these units."

It seems unarguable, further, that the consistent and continual nature of the chronic nursing care required by the hemodialysis treatment regimen is unprecedented among noninstitutionalized populations. Caregivers may be involved, sometimes with the same patients, 4–5 hours, 3 times each week, for years. As mentioned, caregivers develop intense and satisfying relationships with many patients over time. Such long-term relationships may, however, also result in certain stresses for the caregiver. The dialysis patient population presently covers the entire spectrum of socioeconomic and personality types. On occasion, dialysis unit personnel may find themselves forced into a close relationship with a patient whose lifestyle is one of deviance or patently at odds with an individual caregiver's personal values and norms. Such interaction may be stressful and even traumatic for the staff member whose goal is to support and enhance an holistic lifestyle adaptation for each patient.

Burnout.   With the recent advent of complex and highly sophisticated medical treatment modalities, the use of which may produce high stress for the caregiver as well as the receiver, burnout among medical staff is a phenomenon currently receiving much attention. Caregiver burnout has been noted especially in such areas as the intensive care unit, the coronary care unit, and the emergency room, where the use of elaborate and highly advanced technological equipment is becoming the rule rather than the exception. Medical staff burnout seems to be a type of physical, psychosocial, professional exhaustion with the caregiver experiencing such symptoms as fatigue, apathy, depression, and lack of real concern for patients or professional activities. The condition frequently leads to an effect on staff members who terminate their present medical-care activities, sometimes temporarily but also sometimes permanently.

Many of the hemodialysis unit staff members interviewed in the study reported that they had either directly experienced what they perceived to be the burnout phenomenon or they had come very close to it. The following comments provide an overview of their perceptions on the topic. From a staff nurse:

> Burnout is—it's kind of like wanting to give the whole thing up. I went through it really bad last year. As a matter of fact, I talked to a psychologist for a while. I had gotten to the point where I couldn't stand it any longer. It's frustrating to see the patients never get any better. You feel like: what am I really doing? Am I doing any good? If you see them [patients] come in here all dejected and wiped out, physically wiped out, and then they start to improve, then you have a sense of satisfaction knowing well they're feeling better than they did before, you know, if they seem

to be fairly happy. I got really tired of it. People were always getting on me because I got up cranky, but five years is a long time. That's a long time to hang in there. I've worked six days a week in dialysis for the last five and a half years. And a year and a half on call 24 hours a day.

A therapist admitted,

I just was real bummed out for a while. I went through a real bad time, and I just didn't care if it "snowed oats." Or I thought I didn't, but meanwhile the pressure was still building. Yeah, and then gradually I came out of it. I changed shifts and changed my routine a little bit, and rearranged some of the pressures in other parts of my life, and I came out of it, to where I could enjoy my job again. It was very difficult to enjoy it during that period of time. Yeah, I was really burned out. In fact, one morning (I always wake up to an alarm clock) and one morning during this burned out period, it came on with the opening lines of the song "Take this job and shove it." It was so perfect. I just laid there in bed and laughed, because that was exactly how I felt about getting up for work that morning. I mean, I was doing weird stuff, like I'd wander in to work an hour late, and nobody would say anything to me. And it was strange, because I would just wander around, you know.

A unit head nurse described her experience as follows:

I came close to burnout at another unit. And in fact I guess I was burnt out. I wasn't as someone said yesterday, "crispy crisp" yet. But I was almost there. What happened, and I really feel it was helpful, was that I terminated my employment for nearly 8 months. I think a leave of absence is almost the answer to it. Because I think you can take a better look at the forest if you're not underneath the trees. I really think if you can get away and say, "Look, that wasn't so bad and I really loved it. This is what I did best in that job and this is what I really liked." Just gives you a chance to take a breath.

Another head nurse noted,

One of the things I talk to staff about, too, in relationship to chronic care is the high burnout rate of dialysis staff. Sometimes your best dialysis nurse doesn't last but a year because you get too involved, you get too burned out. So the key part of staying in chronic care and staying in dialysis and really being able to administer the best care to the patient is protecting yourself against burnout. I think they have dragged burnout through the mill. But I think it is really true. You get overly involved. And when I see a nurse who is getting too close, I change that nurse to another chronic patient. And say, "Look, you are losing objectivity here." And it has happened. I have had nurses who have gotten too close. The patient has called her at home, all kinds of times in the day or night and been a real drain on that person. I have had staff who have said, "You know, I am dreaming about this patient, worrying about him, and finding that I feel like I am the only one who can run the patient really well." And that is burnout.

Finally, a long-term caregiver described her method for avoiding burnout: I distance myself from dialysis; you just learn—you have coping mechanisms over the years. If I didn't like dialysis, I wouldn't be in it. So you learn what your limits are, then you just kind of stick by it."

While it might be argued that the burnout phenomenon has been somewhat overworked in the extant nursing literature, it remained a phenomenon of notable concern among head nurses, supervisors, and other personnel involved with long-term care of the maintenance dialysis patient.

Informal Staff–Patient Relationships

It has been noted in the literature that "intense relationships between staff and patients are likely to develop"[11] in the dialysis unit because caregivers work with a particular, sometimes small, group of patients over a long period. These relationships have been related to such health–illness associated factors as patient compliance with treatment regimen[10] and overall adaptation to ESRD and hemodialysis.[12] The relationships, however, often take on a character beyond the more professional, instrumental patient–nurse interaction. Kagan suggests that an attachment occurs between a person and a close or intimate companion. She notes, "Hemodialysis nurses are frequently regarded as intimate companions of the patient. They are usually present many times a week and not only provide continuity in the patient's care but also are genuine 'friends' of the patient."[13] Kagen also points out the need of any chronically ill patient for a caring, interested person to listen to their concerns. Anger and Anger specify that a meaningful and caring relationship with a hemodialysis unit staff member may give a dialysis patient a reason to continue living, "since it becomes difficult to want to die if you feel you exist and have meaning to someone else.[14]

In the present study, caregivers reported that some intense staff–patient relationships did indeed develop over time, the dimensions of which varied with particular individuals and at particular intervals during the interaction. As one nurse described it,

You see some of these patients so consistently, day in and day out, over the years that you really do get close to them. At first when you go in to dialysis nursing, you're all wrapped up in the machine and you're doing that and it's really exciting. But that passes very quickly and then the nursing gets into an emotional and social interaction with the patients. You talk a lot on the unit and even sometimes talk with some patients outside. I worked with a nurse who married a dialysis patient. I think most of the socialization, though, occurs with a renal background of some sort. They [patients and caregivers] might meet at Kidney Foundation meetings or there might be a social function that is supporting a Kidney Foundation cause and they get together. I haven't seen it too much on a social, "Hey, let's go out and get a drink" type of thing, because you really can't share the same type of entertainment as far as eating and drinking. But I've seen nurses socializing with families of patients. A great example is the "Smiths." This is a case where you stay in contact and they call you, you know. They're having a party, and they invite you there.

Another nurse commented that she felt it was impossible not to go beyond a purely professional relationship with dialysis patients:

> You work 4 to 5 hours a day, 3 days a week, with the same people over and over. I know as much about most of these patients' personal problems as I do about what goes on in my own family. I spend more time with them than I do with anyone else in any working day. You can't help but be involved!

A head nurse, in discussing why staff often attend funerals of dialysis patients, commented,

> I think it is very common that you go to a patient's funeral. I mean, it's usually something you do for a friend. I think that you develop a relationship with that person. It's a long-term relationship. I think in dialysis you have that long-term consistent contact with somebody, that they become your friend, in essence.

The respondent also denied that such friendships interesected with the caregiving relationship:

> My personal feeling is that it didn't interfere with my work. I felt that if I was dying or if I was being very sick, I think it would comfort me to know that somebody that cared for me as a person, besides just being a patient, was taking care of me. And I felt that I could contribute more to the care knowing the person as a person rather than just a patient. I don't think it really interfered. Sometimes I guess it's harder to be able to be honest with that person, because you know them so well that a lot of the statements or things that you have to tell them is going to hurt them and that might make it a little harder emotionally. But, and then again, if it was the other way around, I'd rather hear it from a friend.

The degree of emotional and/or social involvement varied with individual caregiver–patient situations and sometimes changed over time, as has been explained. One nurse commented that at present she did get very personally involved with patients during their time in the dialysis unit, listening to and sharing their problems, but at this time wished to keep the interaction in the professional health-care setting. She related a previous type of interaction, which she felt she could no longer continue:

> Where I used to work some of my patients had my phone number, and they would call me if they ran into problems at home, if they got sick, and a few of them were home patients, so if they ran into any trouble they could always call me at home. If they had trouble with their machines or things like that, or if they got home from the unit and they got sick or they didn't understand their diet. But I was getting real burned out with that.

Many of the hemodialysis caregivers commented that while informal or social relationships with patients could at times be positive, occasionally conflicts occurred relative to the staff member's professional role as caregiver and personal role as friend or confidante. Of her involvement with patients, one therapist reported,

It's difficult to be their [the patients'] friend and maybe even talk on the phone or go have pizza and then the next day you see them [patients] in the unit and you have to be the "heavy" and they don't want to hear it—they figure they can tell you what to do.

Another caregiver admitted to ambivalence about her own nurse–patient relationships: "I do tend to get very involved, but I shouldn't. It's really not such a good idea." She added, "You know, there is a very fine line between too close and not close enough."

In terms of maintaining the "fine line" mentioned above, one head nurse stated,

As I see it, you have to maintain some kind of professional distance from the patients. But you see the same people over and over, so many times, that it is difficult to maintain that fine line of involvement and professionalism. Often it kind of tends to digress into a social relationship. Then you have to come back to a focus on that person's health or illness. You have to set limits.

Some caregivers expressed less than positive opinions about intensive staff–patient relationships in dialysis nursing. One nurse advocated the concept of frequent staff–patient rotation rather than primary nursing:

This [rotation] helps eliminate some of the [patient] dependency. I think it gets very boring for the staff dialyzing the same patients three times a week, week in and week out. And there's a chance for something new to be learned from each patient. The rotation idea is very, very good and it avoids the problem of patients and staff getting too attached.

In terms of patient–staff relationships developing outside of the caregiving situation, one head nurse expressed a decidedly negative opinion. Her rationale was as follows:

I don't think we should develop social kinds of relationships with out patients. I think that's not a terribly healthy relationship to have, especially in an ongoing kind of care-giving process. There are some patients who are extremely well who you do see outside the unit; they will speak to you on a social level. They'll ask you how your family is and we'll ask them how their family is. I think to that extent, it might be that there's a healthy element to it, because you've gotten to be part of them and they've gotten to be part of you, and it's nice—it's like seeing an old friend again. However, that relationship changes in the event that that individual becomes very ill. You have to become a nurse again, not a friend.

COPING WITH DEATH

Perhaps the most sensitive and painful problem with which the dialysis caregiver must cope is the loss of a patient. Caregivers report that while the degree of loss may vary with such factors as the patient's age, length of

time in the unit, severity of illness, prognosis, and quality of life, the death
of a renal patient is often emotionally draining. Sometimes facing a patient
death may stimulate in the caregiver a degree of anxiety and/or uncertainty
about his or her own termination of life. Kaplan De-Nour and Czackes point
out that working with the condition of "machine dependent continuation
of life" may evoke a number of differing emotional reactions in hemodialysis
team membes.[15] Gelfman and Wilson, in discussing the emotional reactions
in a renal unit, comment that psychiatrists must consider that fear of death
may be a greater dynamic among both family members and caregivers than
it is for the patient.[16]

In discussing the situation of a hemodialysis patient who actively chooses
the alternative of death, Anger and Anger note that dialysis staff members
may, on occasion, be "threatened due to their own emotional state or in-
adequacies. They need the patient to go on living in order to confirm or
sanction their own need to exist."[14] Schowalter et al speculate that "a patient's wish for death forces staff members to confront their personal mortality and sometimes a latent self-destructive wish.[17] McKegney and Lange,
in discussing a patient's conscious decision to terminate dialysis, suggest
that perhaps one of the most difficult facets of the situation for hospital
caregivers is "the staff's obligation to continue to care for the patient as he
slowly dies of uremia."[18]

Dialysis caregivers in the study have reported varied responses to a
question about attitudes toward death and their coping behavior, as the
following data excerpts indicate: "It's very difficult to deal with death. With
some of the patients it takes me a while to really realize that the patient
has passed. I guess I just deny, you know. I have cried, but sometimes I do
deny myself the tears that I'd like to shed because I figure, well, I have to
go on. I have to do what I can for those that are still here. So, I try not to
get myself worked up or too emotional over deaths of patients"; "I tried to
put my mind off it when I started; patients' deaths really don't bother me
too much. If a patient gets four or five years out of hemodialysis I feel they
rode their trolley to the end of the line, you know. The longer the patient
is on dialysis, the more I expect their death." A head nurse commented,

By working here in the unit I know that one day they will die and I've learned
to put up with that. But what's hard is when you're dealing with a healthy kid, you
know, a healthy patient, and eventually they just die of some other complications.
I just had a patient that I trained who died. That was just a couple of months ago.
And it hurt, you know. I tried to avoid the whole issue, and eventually was able to
talk with the wife about a month after. I saw Mr. G. at his viewing . . . but I just tried
to shut myself away from it. After a couple of weeks, I was all right.

A staff nurse described her feelings about the loss of younger dialysis patients
this way:

Sometimes I feel very callous, you know. It's like I have developed this shell through the years. It's very frustrating to see these younger people go. You feel like you're breaking your back, doing everything, and then all of a sudden the patient is dead. And you question yourself: "What did I do wrong? Should I have done more?" You don't want to realize that you're only an attendant on an ill person!

A dialysis therapist reported that while she was beginning to get used to death after about 5 years in the field, occasionally a particular patient's expiring would really disturb her. She observed,

It's getting easier for me [to handle death]. I've noticed that the first couple of times it happened to me here, I cried, you know, I felt really bad for a long time. And now, patients that I know fairly well and like, I feel, "Gee, that's too bad, but maybe they're better off." But one patient died that I didn't even realize that I was particularly close to, but I'd known him for a few years here. There was no personal connection, just strictly as a patient of mine for a few years, and he died—I had to stop what I was doing and take a pillow into the bathroom and beat it up. And he wasn't even a particularly popular patient, but for some reason it just bothered me, to see him go like that. He arrested here in the dialysis unit one night, and didn't make it. And I just excused myself and went out and had to beat on something. I think maybe seeing it happen may have made it worse.

Finally, a dialysis unit social worker commented on patient death as a reality-orienting phenomenon for staff:

I think you have to be free to express your sorrow and your concern. I don't think there's any pat answer. I think people express their shock, no differently than anybody expresses shock at an unexpected death. I think it brings the treatment team back to reality because we know we're dealing with terminal patients but obviously we don't see them as terminal. And this brings us back to reality; often, I've heard myself say, as terrible as it may sound, where do you think statistics come from? We think that we're keeping people alive, and we are, but statistically they die, and they are going to die here as well as anywhere else.

Many caregivers focused upon type of death (expected versus unexpected), patient condition, or quality of life in their response: "It is hard for me only if it's sudden. If I didn't expect it to happen. Like last night this young patient had an M.I. and died and it really shook me up. It shouldn't have happened"; "I had one young man that died of a massive embolus on the machine and that was really hard for me to cope with for a while." One staff member discussed the trauma of a sudden death in detail:

I think if you work with somebody very closely, somebody who has been ill and suffering, I think that nurses will accept that death a lot easier than somebody that walks in, has an acute illness for some reason and dies suddenly. I've seen that with a lot of renal patients. Recently, one guy walked in off the street who was as healthy as could be, got sick, and died within 24 hours. And I've never seen this dialysis unit so upset in my life. Whereas other people that we've had here, like Miss

E. and those that lingered and lingered and lingered—everybody was saddened by her death, but I think they realized that she was out of her suffering.

Another respondent focused upon a "lingering" patient's death:

Well, first, I think, how difficult it is depends on whether or not the death is slow or a dragged out thing. We have had over the last couple of years several diabetic patients who did not do well and have died. And that's very difficult for me to deal with because it is a slow process. And very seldom have I ever seen a patient like that who decides not to go on with the treatment even though they are really bad—you get them on the machine for 15 minutes and then they are begging you to take them off. They are brought here for four hours, you know; so you are in a position where you really cannot do anything.

Several dialysis team members focused upon quality of life in their comments and suggested that a patient's death may, on occasion, be a blessing. As one observed,

Well, I have difficulty dealing with death personally. But I think the way I look at it with these patients, and especially the older age categories that we see, I probably don't let it upset me as much as if they were just walking on the street with nothing wrong with them and they up and died. Especially if they've been, say, having a real hard time with their access or there's some other physical thing wrong with them— they are going down hill. Then many times I consider it a blessing. Because I don't like to see them suffering, and I don't think they want to.

Another respondent stated,

Sometimes it's a blessing when they go; they're suffering so bad, when the end comes, that it is a blessing to see them not suffering any more, that God has taken them. So at times I don't feel bad. It's only the unexpected deaths that I really feel bad about. If somebody here would die in the chair. Somebody who is sick, I can cope with that because the disease is progressing and usually gets worse and worse as time goes on. Patient deaths that are unexpected, patients that you wouldn't think would die, those are the ones that are hard.

One nurse admitted to relating patient death to a recent personal family death, commenting that she has lately had more difficulty coping:

I would say in the last year, I have been more affected by patient deaths because my mother died a year ago. And it's funny, before death affected me, yes, but not bad. It really is hard when a patient dies, even if it's one that's been a demon, that's been rotten to me, kicked me, spit at me, you know, even them, since my mother died a year ago, it's really bothered me.

Several caregivers mentioned the desire to "pull back" from patient involvement for a period of time after a particular patient's death. One therapist remarked, "I guess one of the reasons I decided to try not to get so involved with some of the patients was because the first patient whose house I'd ever been to, I was there for a party, died shortly after that and it was the first death I'd experienced." She added, "After you go through the first hurting experience, then you kind of back away. It's for self-protection, I think, because you realize that it is your job and your livelihood, and if you go around and let a lot of the patients' problems interfere with your life, you know, becoming too involved, you're not effective."

Mutual support of staff members was cited by most respondents as their way of coping with patient death. One nurse commented, "I think being in a unit like this is helpful because we can all talk to each other. We can support each other, when we're finding it rough, when somebody is dying, and I think basically our group is our support." A unit social worker added, "Mostly we cope by supporting and talking with each other. It is hard, you know, we know these patients for such a long time. It really can be painful and sad; but usually we get support just by talking with each other." Finally, a hemodialysis and transplant unit head nurse summarized her thoughts this way:

I personally think that dialysis staff deal with death better than any other staff that I've worked with, any other group of nurses, because I think that they realize that end-stage renal disease is a fatal disease, and that probably in their time one of their patients is going to die. But I just think it's the circumstances that make the difference in how the death is coped with.

The attitudes of dialysis caregivers toward patient death lend credence to the thought of Ignace Lepp that, while we are often not unduly upset by the death of elderly parents or grandparents, for whom the end may be expected, it is another matter when death threatens someone of our own age or younger.[19] A nurse expressed her pain on the loss of a very young dialysis patient, exclaiming, "I was so mad. I still am. He died and I couldn't stop it." Such a response cannot but suggest the poignant words of Dylan Thomas, "Do not go gentle into that good night. Rage, rage against the dying of the light."

A CARE-GIVING TYPOLOGY: THE MACHINE-TENDER, THE COUNSELOR, AND THE CONFIDANTE

Within the qualitatively oriented phases of the present research, focused interviews were employed to elicit attitudes and behaviors of long-term dialysis caregivers; from the beginning, it was anticipated that some serendipitous discoveries would emerge. For this purpose the researcher utilized the methodologies articulated by Glaser and Strauss for the "discovery of grounded theory," i.e., "theoretical sampling" and the "constant comparative method of analysis.[20] Through these processes particular concepts or themes that emerged early on in the data-collection were pursued and clarified. (For elaboration on the methodology of Glaser and Strauss as employed in the study, see the appendix.)

A predominant idea or theme that was verbalized frequently in early interviews with staff members was that of the caregiver's degree or type of actual involvement with the patients. As one staff nurse put it, "You have to make a decision when you start this job about just how involved you want to get with the patients. It can be a 24-hour thing if you let it be. Nurses and therapists are all different about how they act with patients. Some get real involved—overinvolved, I think—others don't really want to be involved at all. But you have to decide—it's really up to you." During the interviewing of dialysis staff members, involvement with patients was examined and out of the data a three-fold typology of care-giving emerged. The caregiver types were labelled the "machine-tender," the "counselor," and the "confidante."

"Machine-tender."   Dialysis staff members who were identified as falling within the category labelled "machine-tender" were decidedly the smallest percentage of the group (approximately 20 percent). These caregivers reported themselves to be "technicians" who were primarily concerned with the mechanical and physiological functions involved in the carrying out of the hemodialysis treatment procedure. They generally denied any real involvement with the patient as person and, in fact, any wish for such involvement. Several "machine-tenders" noted that they felt this was their appropriate role and were comfortable with it; one respondent, however, admitted that she was becoming apathetic and was distressed by this fact. She accepted the role of "technician" but was not happy with it. In her words reportedly,

At first you're so busy you really don't get involved with the patients at all and after a while you really don't even want to. Part of it is the rules on this unit. I'm not working as a nurse; I'm supposed to be working as a therapist and actually doing the work of a technician—you just put the patient on and then watch that machine. You just do technical things—after a while you forget that there's even a patient in that chair.

This nurse added a comment to the effect that this was not a good situation either for herself or for the patients, but that she was "stuck in it." Another dialysis staff nurse added that working in the unit "can get very boring because it is very easy to just get wrapped up in the machine. And you often forget you're involved with a patient, especially when you've got a large patient load." Certain other caregivers' comments supported those presented above, as the following example demonstrates.

Although you do get used to it, I think the running of that machine is your primary responsibility. You've got to monitor those patients—you've got to watch that machine. After all, the patient is only here [in the unit] for 4 hours and the most important thing that happens during that 4 hours is that he get his treatment. He can get psychological help or counseling outside the unit. That's not my job in here.

Finally, one dialysis unit head nurse noted that she felt some of her staff members perceived themselves as technicians because the authoritarian system under which her unit functioned did not allow for freedom or creativity. She stated,

> The nurse has no say. We are not included in admissions, who is going to be accepted into our program. When there is no consideration at the start, what is this going to do to the nurses, who are going to be taking care of the patient? It's all chauvinistic and very paternalistic. One of the staff nurses said recently, "Don't ask me anything. I am just the hands that run the machine." And that sort of summarizes it. It is like we are treated as widgets for what we can do in somebody else's plan.

"Counselor."   Responses regarding self-perceived degree of involvement with patients identified approximately half of the hemodialysis staff members interviewed as falling into the category labelled "counselor." Nurses and therapists in this category reported a more holistic approach to the caregiving process in the dialysis unit. Many suggested that while the technical aspect of their job, i.e., the carrying out of the hemodialysis treatment, was important, it could be easily managed, and that overall psychosocial support, e.g., listening and talking to patients, was a more challenging and interesting part of their roles as caregivers. These respondents generally felt that they should and did get to know their patients as persons; get to know their problems and anxieties, and in some ways, relate to them as friends. The staff members described in the "counselor" group, however, admitted that they wished little or no contact with patients or their families away from the dialysis unit; that they wanted, in fact, to "forget" hemodialysis when they are off duty. One therapist in the above group described her care-giving role this way:

> I'd say my work is mostly psychological and supportive. There's very little physical work involved in dialysis. You can get the patients on quickly and do that on "automatic pilot" without any effort. I like to sit and talk to the patients and try to help them with their problems. One thing though is that I do want to leave it at work. I don't want to take my job home with me. I don't think it's good to develop close relationships, friendships, with these patients outside, because it becomes difficult as far as just worrying about their health, being afraid that they might arrest while you're out with them. Seeing them go through things on the machine and their depression is hard to deal with. It bothers me to think that if you were in the back and there was an emergency, you know, well, there's that initial thought, well it might be so and so, and you come out and you go, "Oh, good, it's only so and

so." And you feel guilty about that. That's bad; so I don't think it's wise to get too ~~personally involved with~~ the patients.

Another caregiver stated that while she got quite close to the patients and their problems while at work, the involvement stopped there. She noted, "I've always lived with the philosophy to leave my job where it is—you can't be involved with these people outside the dialysis unit, too." Several dialysis unit staff members focused on a change occurring over time. As one described it, "Sure, at first your job is very technical. It's all new and you have to learn the machine and all. It's very exciting, but that doesn't last very long and you get bored with it. Then as you get used to dialysis you begin to take more time talking to the patients—working with their problems." Another therapist noted that while she liked being involved with the patients during their treatment time, she did not interact with patients or other staff members outside of the unit. She commented, "You have to have a separate private life that's not involved with dialysis. I like to have friendly rapport but nothing on a social basis."

"Confidante." Approximately 30 percent of the hemodialysis unit staff members who served as study respondents could be considered to fit the category of "confidante" in the care-giving typology. Those caregivers reported themselves to be very involved with dialysis patients and occasionally with family members. Most admitted that they did have contact, professional or social, with some patients outside the unit. Phone contact appeared to be more frequent than actual face-to-face interaction.

Caregivers in the "confidante" group validated the idea noted earlier in this chapter that intense relationships between dialysis patients and staff can develop over time, and they supported the observation that frequent (three times a week) caregiver–patient interaction did in fact sometimes lead to friendships that went beyond the territorial confines of the hemodialysis unit. One therapist noted, "You can't see these people—work with them as often as we do—and not get close. You care and you worry about them even when you're off duty. Sometimes they call you and sometimes you call them. You can't just walk away from it." Interestingly, however, several of the confidante-type caregivers admitted that at times patient–staff friendships posed bothersome to serious problems and sometimes interfered with effective care-giving. One nurse commented that such intense relationships could result in a breakdown of the caregiver's authority and thus entail loss of security for both patient and staff in the professional setting.

Several head nurses commented on their perceptions of confidante-type relationships among their staff and patients. One described a particular staff member in her unit in the following way:

R. is an unusual person, very unusual, in that she is so involved with the patients. She has done more with the patients than most people could. She works another job on top of this, but she still managed to find a way to visit a young girl who had a transplant, every day! When the patient was in the hospital for her transplant—and afterwards when she had such a hard time, you know, and then she rejected and they had to remove the kidney—R. went over and even did her hair and said, "Now, you put some makeup on. You look bad."

Another head nurse additionally noted the difficulty for a confidante-type staff member if a patient became very ill or died. As she described it, "Some of the nurses have gotten really close to some of the patients, become friends with them. If that patient deteriorates, which is inevitable, it's really a hard strain. The nurse or therapist really feels the loss."

Finally, one confidante-type therapist discussed her dialysis patient care-involvement in terms of interaction with other staff members as follows: "We talk about dialysis constantly, no matter where we go or what we do, be it a party, a social gathering, it always ends up work." She summarized her self-perceived role as follows:

The doctors feel like we're just here to do what they describe in our job description. They don't realize how much we do give to these patients, and how much we can offer them, if they're willing to respond. In fact, you go a lot beyond your job description really. You get involved and you care. If you want to go by what they put in the job outline, of course, you won't get involved. Who ever pays any attention to that anyway? I mean I'm not saying that I would ever offer the patients psychological counseling, but I was able to refer a patient to someone who was able to help her. And she's doing fine now. And I think from that, you see, how much the personal interaction helps.

[It should be noted that professional caregivers, nurses, paraprofessional caregivers, therapists, and technicians, were included in all three categories. Neither group appeared to be notably more attracted to any particular one of the three categories.]

It is difficult, in conclusion, to present a neat and totally unambiguous definition of the hemodialysis caregiver. Dialysis unit staff members in the study presented a myriad picture in terms of age, educational background and sociocultural experience. Yet, despite these differences, a universal theme of caring undergirds their attitudes and behavior. Caring was at one time fiercely manifested as anger, as frustration, as rage against their patients' consuming illness condition and the life constraints imposed by its associated medical regimen. Caring, at another time, was expressed gently in words of understanding and empathy for the painful illness experiences that were often internalized by the caregiver in the process of therapeutic interaction. The hemodialysis unit staff members repeatedly and forcefully expressed their humanity in describing the caregiving process. The philosophy underlying their attitudes and behaviors is reflected in the classic words of Samuel Johnson (Rasselas):

To live without feeling or exciting sympathy, to be fortunate without adding to the felicity of others, or afflicted without tasting the balm of pity, is a state more gloomy than solitude: it is not retreat, but exclusion from mankind.

REFERENCES

1. Gutch CF, Stoner MH: Review of Hemodialysis for Nurses and Dialysis Personnel. St. Louis, CV Mosby, 1975, pp 1–2
2. Nouwen HJM: The Wounded Healer. Garden City, N.Y., Image Books, 1979, p 82
3. Tractate Sanhedran, as cited in Nouwen HJM: The Wounded Healer. Garden City, New York, Image Books, 1979, p 82
4. Light D, Keller S: Sociology. New York, Alfred A Knopf, 1982, p 334
5. Goodman N, Marx GT: Society Today. New York, CRM/Random House, 1978, p 556
6. Matuszak DF, Sharp NJ: The effects of peer support and sharing among head nurses of dialysis/transplant units. J AANNT, 4(suppl):31–34, 1977, p 32
7. Lane CA, Hawkins A: Managing stress in a hemodialysis unit. J AANNT, 8:36–37, 1981, p 37
8. Lira FT, Mlott SR: A behavioral approach to hemodialysis training. J AANNT 3:180–188, 1976, p 180
9. Kroemeke GT, Nassar T: An evaluation of ward atmosphere in hemodialysis units. J AANNT 7:282–284, 1980, p 283
10. O'Brien ME: Hemodialysis regimen compliance and social environment: A panel analysis. Nursing Research 29:250–255, 1980, p 250
11. Roper E, Raulston A, Cramer D: Attitudinal barriers in dialysis communication. J AANNT 4:179–198, 1977, p 179
12. O'Brien ME: Effective social environment and hemodialysis adaptation: A panel analysis. J Health Soc Behav 21:360–370, 1980, p 364
13. Kagan LW: Renal Disease. New York, McGraw-Hill, 1979, p 190
14. Anger D, Anger DW: Dialysis ambivalence: A matter of life and death. Am J Nurs 76:276–277, 1976, p 277
15. Kaplan De-Nour A, Czaczkes JW: Emotional problems and reactions of the medical team in a chronic hemodialysis unit. Lancet 2,7576:987–991, 1968, p 987
16. Gelfman M, Wilson EJ: Emotional reactions in a renal unit. Comprehensive Psychiatry 13:283–290, 1972, p 283
17. Schowalter JE, Ferholt JB, Mann NM: The adolescent patient's decision to die. Pediatrics 51:97–103, 1973, p 99
18. McKegney FP, Lange P: The decision to no longer live on chronic hemodialysis. Am J Psychiatry 128:47–54, 1971, p 53
19. Lepp I; Death and Its Mysteries. New York, Macmillan, 1968, p 59
20. Glasser B, Strauss A: The Discovery of Grounded Theory. Chicago, Aldine, 1967

─────5─────
The Dialysis Unit as a Social System

One must observe the proper rites. . . . "What is a rite?"
asked the little prince. "These are actions too often neglected,"
said the fox.

Saint-Exupéry
The Little Prince

On entering the hemodialysis unit, one is struck by the casual, almost (though not quite) relaxed atmosphere of the place. After initial adjustment to encountering the "machines," those wonderful, fearful mechanical monsters, through whose myriad tubes and coils the patients' blood is coursing, the social life of the unit is observed to emerge with continually changing scenarios of interchange among the actors. Patients and staff meet in a territory where there are vested as well as mutual interests; caring and conflict may occur in the same exchange. Patients struggle to come to terms with their illness and their total dependence upon medical machinery, sometimes engaging in noncompliant behaviors in order to preserve some small measure of control over their destiny. Caregivers, mindful of their roles as comforters and healers, rage against such violations of the therapeutic regimen, expressing at the same time their frustration and anger at the overwhelming life restrictions imposed upon their patients. Here in the hemodialysis unit where life is both preserved and threatened, many patterns of attitudes and behavior related to survival and life quality are observed among the participants. Each dialysis unit is unique in sociocultural milieu, with its own particular patient and caregiver populations, yet many interactional commonalities may be identified.

Hemodialysis units also vary in terms of size (physical space), decor, location, number of chair-machine units, type of machines, type of dialysate delivery systems, treatment times, and staffing patterns for the provision of patient services. Some hemodialysis units are free-standing commercial facilities, while others are hospital or medical-center based departments providing for acute as well as chronic renal failure conditions.

The dialysis center commonalities were examined in the present study by evaluating the unit as a social system, with specific goals, norms, status roles, rank assignment, beliefs, sentiments, power structure, and sanctions. Loomis has suggested that a social system consists of the interaction of a number of individuals, "whose relations to each other are mutually oriented through the definition and mediation of a pattern of structured and shared symbols and expectations."[1] In the dialysis unit there is found a basic pattern of goals, symbols, and expectations generally shared by both patients and staff. As the following analysis demonstrates, however, disagreement and conflict over specific expectations and behaviors may occur between caregivers and patients, and resolutions at times prove difficult, if not impossible.

The hemodialysis unit possesses both an "internal" and "external" environment. The external environment or boundary—"how the group is brought together"[2] is the physical facility, i.e. dialysis unit, where a treatment procedure for end-stage renal failure is carried out. The internal environment or system—the group behavior that expresses group members' sentiments toward one another—is comprised of the attitudes and behaviors of dialysis unit patients and their caregivers with relation to the hemodialysis procedure and total regimen prescribed for the illness condition.

GOALS OF THE DIALYSIS UNIT

A primary goal of the hemodialysis unit is technological—that of removing bodily waste products through utilization of a machine commonly labeled the artificial kidney. For most patients the procedure is carried out 3 times a week and lasts approximately 3 to 4 hours. Some dialysis unit caregivers reported that they felt their central task was technological—to monitor the patient and the machine and to ensure that an adequate treatment procedure was conducted with as few physical problems for the patient as possible. One nurse-therapist stated, "My main job here is technical— you just put the patient on and then you watch the machine. After a while you forget that there's even a patient in that chair." A dialysis unit supervisor observed,

I find it enormously frustrating trying to motivate people and also to help people that are motivated, to remain so, because the job itself is so highly technical. It's very easy to get overinvolved in technical aspects and either be too busy or just get so regimented that you don't even see the patient as an individual. You don't empathize any more with what he is going through.

A larger proportion of caregivers, however, articulated a more holistic approach to patient well-being and were concerned with overall adaptation to maintenance hemodialysis and the treatment regimen. These caregivers suggested that a secondary but very important goal of the dialysis unit was to promote and facilitate psychosocial as well as physical patient adaptation in terms of activities of daily living and quality of life. One therapist noted,

Initially, I think it [the caregiver role] is more technical because you have to learn patient response to the machine settings and things like that. You're very uncomfortable. But after you've gained more experience, your job becomes more one of psychosocial support in the sense that you become involved with the patients emotionally. You try to listen to their problems, help solve them, work through different things.

A distinction was noted by several caregivers between acute and chronic units, acute units being envisioned as more technically oriented, as they are frequently involved with more seriously ill patients who often are admitted in crisis.

The goals of the hemodialysis unit are reinforced by the normative structure articulated in federal regulations mandating appropriate patient services and standards of care. Most units provide structured orientation classes for all new personnel involved in caregiving, as well as periodic inservice education sessions for continuing staff. Overall, it appears that more holistic patient adaptation goals are supported by both professional personnel and patients. Although some study patients expressed concern about the conduct of the dialysis procedure and the technical skills of the caregivers, most voiced a concern that those carrying out the long-term dialysis treatment "care" for them as persons. As one long-term dialysis patient noted, "It really means something to know that these people around here care for you. Without that, you can hardly make it."

THE TECHNICAL ENVIRONMENT

Hemodialysis machines or "artificial kidneys" come in various shapes sizes, and vintages with new and/or refined equipment being produced each day in our age of scientific and technological explosion.

The most commonly used dialyzers are the coil, the flat plate, and the hollow fiber kidney. The hemodialysis machine generally receives its dialysate through a central supply system located in the unit. Two basic types of systems are the batch system and the proportioning system. The batch system "involves the preparation of a large amount (100 to 120 quarts) of dialysate by mixing the concentrated commercially prepared chemicals with large amounts of purified water." This system requires a great amount of space. The proportioning system is so designed that "continuous automatic mixing of concentrated chemicals with purified water goes on during treatment."[3] An advantage of the latter delivery system is that it requires less physical space in the dialysis unit for storing the dialysate.

Most hemodialysis patients know their machines quite well and often become "attached" to a particular "artifical kidney" in the unit. One 42-year-old woman who had been undergoing dialysis for approximately 11 years and receiving treatments at the same dialysis unit during that time, complained that she had recently been having many physicial problems during her treatment sessions. The patient attributed this situation to the fact that "her machine" had been "taken away from her" and given to a patient who required isolation. She stated, "This machine [the one she was using that day] doesn't do right for me. That machine [her former one] really ran me good and took the fluid off, but these old machines are no good and I just build up with fluids if they use these on me." A considerable number of patients reported a change or proposed change in their unit's treatment equipment but most reported being more comfortable with the technology presently in use. It appeared that a certain security develops over time in regard to the machine, and the potential for change may pose threats to one's continued treatment adequacy and perhaps even survival itself.

THE PHYSICAL ENVIRONMENT

As with types of units, the physical environments of the hemodialysis units vary to some degree. Generally in the larger units (30–40 patient stations) machines and chairs are arranged in parallel rows in one long treatment room for ease of observation by the staff. Some smaller centers (5–10 patient stations) place the patients in a semicircular type of arrangement. The decor is primarily functional, but certain facilities have incorporated a few "homey" touches such as brightly colored drapes and plants hanging in wicker baskets from the ceiling. Most dialysis units also have functionally designed patient waiting rooms, staff lounges and centrally located nurses' stations.

Despite the attempt to provide a pleasing ambiance in the dialysis units, patients frequently reported being "overwhelmed" and "terrified" when they entered the treatment room for the first time. One patient stated that "all of the machines with blood in them made me feel like fainting," and said she was "scared to death of what it [the treatment] would feel like." Some family members also reported that the appearance of the dialysis unit was very frightening at first. The mother of one patient reported that while she did go to visit the unit when her daughter first began dialysis, it caused her much distress to do so. She said, "It was just morbid—I hated to go on that unit, but she was on it, so I did—I never did tell my daughter that, though."

After the initial adjustment (and sometimes shock) wore off, however, most patients and family members conceded that they became comfortable with the physical milieu of the dialysis unit. One patient said of his unit, "It gets to be like your second home, so you'd better like it." He added, "Besides, your life depends on this place." Perhaps "beauty," or in this case at least "acceptance," is in the mind's eye of the beholder.

TERRITORIALITY—THE CHAIR

It was found in the present research, and repeatedly validated by caregivers, that dialysis patients tend to establish their "territory" within a unit. The investigator could return to a dialysis unit after several years and expect to find a patient in the same location as previously. Some patients were reported to have their "own" chairs and pillows, and any major change or reassignment of seating was reported to wreak havoc among a particular shift treatment group. One dialysis patient put it this way:

The chairs and where you sit are a real big thing around here. They're redecorating and remodeling the place and they're going to have to move some people and you're going to see a real uproar. People want their same place and same chairs every time and get really upset if they get changed. The new chairs are different from the old ones and sometimes people say that can't take the treatment in a different chair.

A caregiver added, "It's incredible! We changed into new chairs and you really would have thought that we were cutting off their [the patients'] right arm. You just don't change their chair. It's a matter of security." A staff nurse in another dialysis unit reported, "We have this patient who always insists on the same place; if he's in that place he has a really good treatment. If he's in another [place] he says he always get sick. He complains at the end of the treatment that we made him sick."

One might speculate as to several reasons for such territoriality among the dialysis patient group. The first may be, as articulated by one of the caregivers above, security, or the fact that the patient has "survived" in that chair and location over time. Another reason for territorial preference may relate to friendship ties with other patients. Who sits next to whom is often a way of describing or locating an individual within a unit. Finally, it was reported that patients form close relationships with particular caregivers, and may request to be treated in an area for which their usual or favorite therapist is responsible. Overall, patients appear to want to maintain the therapeutic treatment situation that has "worked" for them in the past. As one young male hemodialysis patient commented, when discussing the possibility of another dialysis modality or of kidney transplant, "Things have been going okay for me so far so my attitude is, 'Let's not rock the boat'."

THE SOCIAL ENVIRONMENT

Social interaction within dialysis units may vary according to the size of unit, shift time, and openness and attitudes of the professional staff. One patient complained that the atmosphere among the patient group on his shift had changed with the arrival of a new head nurse in the unit. He complained with the "girl in charge makes it bad on us and on the therapists. You don't have your own technician any more. They [the therapists] used to stand around and talk to you—you had somebody to cut up with and pass the time away with when you were on the machine. "Now," he added, "everybody is uptight around here."

Another factor that is reported to have lessened the interaction between patient and patient and between patient and staff is the introduction in some units of individual TV sets at each chair-machine location. A therapist complained angrily that with the advent of individual TV sets in her unit patients no longer spent time in verbalizing their concerns and problems to the staff. She stated, "I would love to tear them [TV sets] all out. The patients come in and they turn on that TV and just ignore life. And I think they're withdrawn and have problems that they do not express and I think that's terrible."

Several other caregivers reported their perception that the presence of television sets at each patient station was detrimental to patient–patient and patient–staff interaction. Although such a negative interpretation of the use of this recreational technology during dialysis may have impact for the social interactional milieu of the unit, a positive benefit may also be cited. A temporary denial or escape from problems may be healthy in allowing the patient to regroup and go on about the business of living. As one of the study patients expressed it, "You just can't make it if all you do is think about this thing [dialysis]. You've got to forget yourself sometimes and just think about the rest of the world."

Coming to the hemodialysis unit was found for some patients, however, to be viewed as a social activity in and of itself. One patient said, "The dialysis unit is like your second family," and some patients were found to remain at a particular unit some distance from their homes even when a new center had opened at a much closer location, in order to "be with their friends." One elderly patient reported that she continued to travel from her home in the city to a suburban unit for the sake of her cab driver, who always provided her transportation, and to "see her friends." Another patient's daughter noted, "I told my mother about the new dialysis unit close to our home but she says she'd rather be out there [her original dialysis unit] with her friends."

Since the advent of PL 96-203, making dialysis available and accessible to most ESRD patients, regardless of such factors as age, socioeconomic status, or cultural background, the patient population has broadened considerably. Thus patient–patient interaction takes on an interesting dimension as dialysis patients from vastly different ethnic and socioeconomic status groups come together in the treatment settings. Frequently, cultural and social class lines blur in the common experience of renal disease and its treatment, and close friendships develop across class and cultural lines.

In one dialysis unit discussed in the literature, music therapy was used as a "special tool to bring about positive change in social growth" among patients.[4] It was believed by the staff that such musical interaction provided a less threatening vehicle for patients' expressions of their feelings and helped build a sense of group solidarity and camaraderie among the patient population. In the present research the number and intensity of patient-to-patient relationships varied depending partially upon length of time on dialysis and adaptive type.

STAFF–PATIENT INTERACTION

Social interaction among patients and caregivers in the hemodialysis unit generally was observed to be informal and congenial. First names or "nicknames" are the usual form of address, and touch is utilized mutually by patients and staff, in friendly gestures. The patting of an arm or placing an arm around a shoulder is not unusual. Obviously, individual personalities and degrees of acquaintance vary in the interactional situations, but more formal types of behavior appear to give way to informality rather more quickly in most cases. This frequently occurs in caregiving situations where the patients' feelings and even intimate bodily functions may become involved in the interaction. Friendships and "play" kinship relationships also develop in such a milieu. On one visit to a dialysis unit a female patient was heard to call one of the male therapists her "adopted son" and his greeting to her was "Hello, adopted mother." The patient noted proudly, "He's my heart. He always takes care of me."

Interaction between professional and paraprofessional dialysis unit staff members and patients was initiated alternately by staff or patients and seemed to depend somewhat upon one or more variables, such as subject of the interaction, patient seniority (years coming to the unit), staff seniority (years of working in the unit), and individual personality of the actor. It was frequently observed that "long-timers" (patients who had been receiving dialysis treatments at a center for several years) appeared to possess a sense of power or control in social interactions with staff and were more comfortable in initiating purely social discussions. One long-term patient who had been coming to a particular unit for over 5 years exemplified this fact. The patient, a middle-aged man, instead of responding to the change nurse's greeting of "How are you?" as he arrived in the unit, asked her "How are *you?*" and added, "Did you have a date last night? There was a full moon out there!" Sometimes patients stepped out of the patient role in interacting with the staff on a social topic and resumed the role when the actual hemodialysis procedure was initiated. Similar behavior was observed among the caregivers. One nurse noted, "After working with some of these people 3 times a week for 3, 4, or 5 years, you can't just be formal and play nurse–patient—you have to show some humanity and act like a normal person." Caregivers also reported, however, that close patient–staff involvement, while sometimes positive, could cause problems in terms of the caregiver maintaining his or her position of authority in the unit. As one therapist stated, "It's like walking a very fine line—you have to be involved and caring but also not get over-emotional and let the patient control you."

THE WAITING ROOM

Much dialysis patient-to-patient interaction occurs in the patient waiting rooms located in most hemodialysis facilities. Patients may have from a few minutes to an hour or more to wait for their treatment to be initiated, and it was observed that some patients arrive at their unit one or two hours early just to socialize in the waiting room.

In evaluating the patient-to-patient interaction, one finds that there is variation depending upon the shift time and patient group. In one unit waiting room, the patient group clearly presented itself as a "clique," with patients and family members seeming to know each other quite well. Although the patients present represented a variety of ages, sexes, races, and socioeconomic status groups, conversation flowed quite comfortably. One gets the impression that ESRD and maintenance hemodialysis may constitute an equalizing force in presenting a commonality to which persons from various cultural and/or economic groups can relate. During this particular waiting room observational period, a number of complaints about the unit—e.g., the air conditioning being kept on too high ("We'll have to start wearing thermal underwear")—were expressed. Treatment "chairs" were also a hot topic and loud complaints were voiced about a notice that had been posted in the unit stating that chairs could no longer be rearranged by each shift at the patient's request. The staff were also discussed, and individual caregivers' ability to carry out the treatment procedure were evaluated. One patient stated, "Some of them [therapists] can run me right down to my target weight and some can hardly get a kilo off." The group seemed united against the staff, and also appeared somewhat closed, as an "elite club." This latter aspect of the patient group composition was manifested in the criticizing of a new female patient who had recently begun to receive treatment at the unit. She was faulted strongly for not coming into the waiting room when she arrived but rather standing in the hall until it was time to enter the treatment room. Obviously, this patient was violating the group norms. As another female patient put it, "What's wrong with her? Does she think she's better than we are?"

Several dialysis patients reported that they enjoyed interacting with fellow patients in the waiting room because they could talk about "dialysis" and "being sick" and their companions would understand. One respondent said, "After awhile your family and friends get tired of hearing all this sickness stuff, and besides, they don't really understand because they're not on dialysis." Patients who were more "senior" in the study group reported that they sometimes tried to help new arrivals; and the wife of a patient who had been on dialysis for 11 years noted that her husband frequently talked to new patients in the waiting room "to encourage them to stay on the machine and not to die."

THE BREAK ROOM

The break room or staff lounge, in distinction to the waiting room where patients may privately discuss the staff, is a place where staff may privately discuss the patients. It is a place where caregivers may go to regroup and to interact for mutual support, talking or just relaxing from the stresses of the treatment room, or "floor," as it is sometimes called. Because of the tension of constantly supervising a potentially life-threatening procedure, staff were often observed during break periods to be either quiet and subdued or overly jovial and joking often about what might appear to be rather serious matters. One unit supervisor admitted that when staff members get together informally they often do talk about "work" and about patient situations in a humorous vein. She noted, however, that "it's not ever done in a cruel way, but a lot of times we laugh about work. We just joke about people's mannerisms and little things—it's a way of coping with dealing with patients who inevitably are going to die one day. Otherwise we'd be gloomy all the time." This caregiver hastened to add that staff members, even while joking, cared very deeply for their patients, and stated, "We'd do anything for them."

It has been observed in studies of crisis-oriented or high-stress medical care situations that humor is a device frequently utilized by professional

staff members to help them reduce tension and cope more easily with burdensome responsibilities. In hemodialysis care, while crisis situations are
becoming less frequent as the technology and competence in patient management improves, the potential for possibly life-threatening situations remains; however, these threats to life inherent in the life-saving procedure
itself cannot be safely dismissed for long from the consciousness of either
patient or caregiver.

POWER—AUTHORITY AND DECISION MAKING
IN THE UNIT

In general, power in the hemodialysis unit might be said to reside within
the professional staff—the administrators and caregivers—whose roles and
functions are assigned in clearly delineated patterns of division of labor.
Although these structured behavior patterns may vary to some degree from
unit to unit, a basic hierarchical ranking of a set of status roles includes
Medical Chief (physician), Administrator, Head Nurse, Shift Supervisors, Staff
Nurses, Therapists, Machine Technicians, and Unit Clerks. (Maintenance
personnel may or may not be directly employed by the dialysis unit.) Although the physician's physical presence is not continual on the unit, he
or she is clearly the symbolic leader or authority figure who orchestrates
the total medical management of the patient. As is still usual in American
medical contexts, in the physician role resides the power to give or rescind
orders regarding the therapeutic regimen—the power, in fact, to initiate or
terminate the treatment itself. Both staff members and patients are required
to look to the physician for direction, support, and positive reenforcement
of their health-care related behaviors. The degree to which such direction
and support are both sought and received varies with individual staff–physician and patient–physician relationships. Overall, however, the physician
remains the dialysis unit's dominant figure.

It is interesting to note, however, that although status, function, and
salary differ for the above roles, several caregivers reported that the shift
supervisor, staff nurses, and paraprofessional therapists all interacted on a
similar professional level, and one therapist noted, "We're all the same."
Knowledge about and skill in carrying out the hemodialysis treatment procedure appeared to have a somewhat equalizing effect among caregivers.
Occasionally status role-incongruence was observed when an experienced
and skilled paraprofessional dialysis therapist was assigned as shift supervisor,
while a less experienced registered nurse performed as a member of the
direct care-giving staff. Such role incongruence seemed to be of little concern
to the caregivers involved. An explanation for the fact that serious staff ten-
sions were not evidenced in study interviews might relate to the diminished

role-strain associated with the one-to-one, patient–caregiver assignments in the dialysis unit. Although staff members sometimes worked in pairs in particular sections of the unit, each caregiver is generlly responsible for "putting on" and "taking off" his or her own assigned patients, and thus retains a degree of personal autonomy and authority regarding the care-giving activity.

Attitudes toward dialysis patients' self-care and self-responsibility were found to vary among staff members, but on the whole it was considered that patients should be knowledgeable about their condition and treatment, and self-responsibility was encouraged. One nurse related her frustration with a patient's continued noncompliance and stated, "He knows more about his condition and his body than I do—he's been living with this for 10 years and it's his responsibility to take care of himself, not mine. It's crazy for me to keep lecturing him about what he should do—he knows!" One physician, however, expressed a concern, stating that, "It can be difficult dealing with chronic renal patients because they sometimes think they know more than you do." The physician added that this sometimes led to conflicting patient–physician attitudes in regard to compliance with the treatment regimen, and noted that sometimes dangerous noncomplaint patient behavior was the result.

The degree to which individual dialysis patients actually do exercise self-responsibility varies notably and will be discussed in more detail in later chapters. Generally, most patients appear to want some control over the treatment itself, telling the staff members at what speed to run their machines and watching the coil and lines during the procedure. Some conflicts have been reported, however, between patient and staff over this aspect of the treatment. One woman complained, "I told them they couldn't run me at 300 today [300 cc per minute] but they did anyway and now I'm paying the price."

As a rule, however, most caregivers not only allow but in fact encourage both patient self-responsibility and self-care during the period when the hemodialysis treatment is being conducted. A therapist describing a long-term patient noted, "She always tells us how to run her and she'll watch her own machine. She's been at this a long time." Certain dialysis units have self-care shifts on which patients are taught to initiate and carry out the treatment procedure with only minimal assistance from the staff; there are also certain centers where all of the patients practice such self-responsibility. Self-care dialysis is described by Gutch and Stoner as "a compromise between center and home dialysis."[5] They note that patients "are responsible for the major portion of the setup and cleanup of equipment and as much self-monitoring of the procedure as possible.[5] An evaluation of the self-care dialysis program in one Veterans' Administration hospital found that self-care patients "take better care of themselves and have more positive self-images. They take an interest in their treatment and are knowledgeable about their disease. They don't have the problems of dietary indiscretion that is so prevalent in our in-center population.[6]

One self-care patient in the present study proudly reported, "I have been doing so much better with this self-care way of treatment—you feel a sense of freedom when you do it for yourself. Sometimes, if I'm late or if I don't feel too good they do it for me, but I miss doing it myself. Now I'm even takin' myself off." Another patient observed, "I run the machine myself—that way I don't have to come in here and wait for somebody else to do me. Sometimes they [the therapists] turn the machine way up and run you off too fast. Then you go home with no strength at all. They think that's the best way, but really everybody should learn to do themselves. If you don't run yourself over 200, you don't lose that little strength you have."

Conflicts occasionally surface regarding a patient shift reassignment or therapist change, and patients tend to be very conscious of "who sticks who" (what caregiver initiates the procedure). Some dialysis patients reported refusing to let a new technician "practice" on them. One male patient, however, noted that he had a "real good" fistula and proudly added, "All the new ones practice on me—I break them in."

In general, it was found that authority and decision making were shared in most hemodialysis units not only among various members of the professional and paraprofessional staff but also between patients and staff. The atmosphere appeared one of "peaceful truce," or "let's try to live together," as may often be found in a long-term relationship not of one's choosing.

THE MACHINE—GIFT OR PUNISHMENT

During the decade of the 1960s hemodialysis was still considered a very special and unique treatment procedure. It was reserved for the "elect," or those patients deemed worthy by medical center selection committees, or "God committees," as they were sometimes called. At that time resources were scarce and treatment with the artificial kidney machine was prized and highly sought after by many patients with end-stage renal disease. Even today, when there is an adequate supply of technological equipment and dialysis facilities to meet the needs of most urban areas in the United States, patients still express their gratitude for the "machine." This sentiment is manifested by expressions such as, "Without that machine I'd be long gone"; "It [the machine] and God have brought me through"; "It's the best thing medical science ever did—to come up with the artificial kidney—it's a real gift."

Ambivalence toward the machine and the treatment procedure was noted among study patients, however. Some expressed their anger directly and verbally, venting their frustration at having to be "tied to the thing" and "not be able to do what you want, when you want." One patient stated,

"Sometimes you feel like it [the machine] runs your life," and added, "After awhile there are days when you just don't think you can get up and come in to the unit and face the thing [machine] one more time. Sometimes, I really ask why did this have to happen to me?" Such an attitude of anger at the machine or viewing the treatment as an unwelcome burden or punishment was also reported by caregivers to be manifest fairly frequently among their patient groups. One therapist commented, "Patients will do anything to get off the machine a few minutes early—some are very creative and manipulative." She added, "They [the patients] really watch their time—and they watch other patients' time. There are patients who get really upset because somebody got on after them but gets off first, but they don't realize that there are varying problems and varying times for dialysis." She continued,

> Don't you be two minutes late. They [the patients] are ready. When it's time to come off they know. They try to con you—it starts about 15 minutes before time to come off. There are patients who will stay through the whole treatment and not say a word. But when it's time to come off, they better be off. I have made a habit of letting the patient tell me what time they went on the machine. I look at the clock too, though. Some will say that they went on according to the time that you put in that first needle. Others will say when you started up the machine; others will put it when you gave them their heparin. If they can cut two seconds, they will. In a sense they act like it's a punishment.

One dialysis center head nurse analyzed the patients' "machine attitude" as relating directly to the present availability of "artificial kidneys" and dialysis centers. She noted that now patients have more power and control over their conditions and treatment regimen and added,

> Even though they [the patients] have their own units, their own chairs, just knowing the availability of dialysis and knowing that it is out there and not hard to get makes them confident. If there's a very short amount of one thing then you tend to hold it as dear. But when you know that it's much larger and there at your grasp, then it's easy for you to say, "Okay, fine, I've had enough of this, I can call the shots." It gives them some sort of independence.

The themes of "gift" and "punishment" were found consistently in study patients' responses. The trend, however, was toward viewing the machine as a life-saving gift for which gratitude rather than anger was the controlling attitudinal response.

THE RITUALS—GOING ON AND COMING OFF

Hemodialysis for in-center patients may be initiated either by the patient, if he or she is practicing self-care, or by a therapist. The arterial and venous blood lines are connected to the appropriate sides of the cannula or to the needles in the appropriate positions in the dilated vessels close to the AV fistula (or graft).[7] The procedure may involve more or less physical pain depending on such variables as position of graft or fistula, toughness of skin, skill of the person initiating the procedure, and the individual's pain threshold. There is, however, for many patients, some degree of stress or anxiety during the initiation period, as well as during the "coming off" time when blood currently in the dialyzer is returned to the patient's circulatory system and the artificial kidney is disconnected. Some dialysis patients were observed to engage in very detailed and specific rituals during the "going on" and "coming off" periods of the treatment procedure.

These patient rituals before "going on" involved such behaviors as the setting up of one's chair (with pads, pillows and sheets) in a most meticulous manner; the placing of food, extra clothing, and reading material exactly in the same spot before each treatment; and movement of one's chair several inches to the right or left to be in the exactly "correct" location for the procedure. The degree to which these rituals were carried out varied among the patient group. One patient was observed to spend almost 15 minutes prior to treatment initiation in the organizing of her chair and personal belongings, which included clothes (house-slippers, sweater), blanket, food, reading materials, and sewing. A dialysis unit staff member noted that this patient had once absolutely refused treatment when a new head nurse had interrupted and, in fact, attempted to put a stop to the patient's ritualistic behavior. The staff member commented, "She [the patient] really went into a tailspin, got all hysterical and said she didn't want her dialysis." The therapist added, "I think patients have a right to do their own thing—after all it's their body going through this, not ours." A hemodialysis unit head nurse, in discussing such rituals, remembered observing a number of patients who exhibited what she described as "very strongly compulsive behavior." She noted, "They would come in with a bag of equipment that they would arrange before dialysis. It may be a blanket, their lunch, a radio, what they were going to read, and it all had to be in exactly the same place." This caregiver's interpretation of such behavior was that of the patient's need for control. She said, "It is really the only thing they [the patients] have control of during the treatment. They have no choice as to who will put them on or whether they will be first on the machine or last on the machine. It is basically the only thing that they can control during that time period."

Some "long-timers" (long-term patients) were pointed out as being especially ritualistic. One caregiver graphically described a male patient who always wanted his coffee with "such and such number of sugars" immediately after he was put on the machine. The caregiver noted, "I mean his blood pressure could be crashing to zero and he'd still have to have his exact same coffee."

Another staff member commented that while most of the patients performed their own rituals before "going on," occasionally a family member did it for them. She ventured her opinion that the ritualistic behavior was a factor increasing security—"the fact that one survived and did well if one did it this way"—and also noted that patients practiced ritual "coming off" behaviors as well as "going on" activities. She asserted, "Some patients leave just as ritualistically as they came," and described these "coming off" behaviors as including the preparation of personal belongings for leaving; the straightening and organizing of one's chair and supplies; and the cessation of all reading, talking, or watching TV in order to be "ready" to come off on time. As a caregiver insightfully observed,

> I think there is a tremendous need for ritual among these patients—it has to do with security and control. You know, there is something very mindless about just walking in, sitting down, and plunking out your arm. The one thing that they [the patients] can participate in is the "preparation" and the "leaving," unless they are self-care patients.

She added, "I think it's a very comforting thing for the staff too—and they have their own rituals which give them security. They don't like those disturbed."

It might be speculated that in a setting such as the hemodialysis unit, where technological procedures involving life and death are dealt with so routinely, these ritual manifestations of the need for security and sameness are perhaps simply reflections of the patients' need to express their humanity and individuality, as described so poignantly by Saint-Exupéry in *The Little Prince,* "One must observe the proper rites."[8]

THE LANGUAGE

Coupled with the new and complex technological equipment and procedures encountered by patients entering a hemodialysis unit for the first time, is a new and complex language that must be learned in order to facilitate communication among and between the actors. This language, as Loomis suggests, is one of the "shared symbols" utilized to express understanding and group identity by the members of a social system.[1] The dialysis unit language relates not only to the technological procedure of hemodialysis and physical attributes of the artificial kidney, but to the physiological and psychosocial characteristics of the dialysis patient as well. The new hemodialysis patient together with his or her significant others must now learn to discuss and understand the patient's bodily condition, utilizing such words as "chemistries" (sodium, potassium, creatinine, calcium, phosphorous), "crit" (hematocrit), "clotting time" (blood clotting time), "target weight," "overload," "retention" (fluid retention), and others. In terms of the procedure itself, labels such as "access," "shunt," "fistula" or "graft," and "bruit" must be understood, and phrases such as a "good sticker" who "doesn't run me too high" have particular meaning to the patient–therapist relationship. If a dialysis patient has taken on and retained too much fluid since the previous treatment, a usual comment to a caregiver might be "I need to be run high today because I'm 4 kilos over target," thus resulting in mutual understanding of the patient's condition and treatment need. If a patient should call out "my coil is ruptured" or "there's air in the line" during the procedure, it is understood that a medical emergency exists and it must be responded to as such. One patient who was beginning to learn about and initiate self-care described her experience with the "foreign" language this way: "At first it was really scary—you don't understand. There are so many things to remember. We had classes every day on words like 'dialysate' and stuff like that."

Most dialysis patients appeared to learn and internalize hemodialysis language very quickly, perhaps because of its importance for their care and even, at times, their very survival. The language was frequently heard in patient-to-patient communication, seemingly flaunted by the more sophisticated as a symbol of their knowledge of and control over this sometimes "uncontrollable" illness condition and its treatment modality. Caregivers (especially new therapists) in the hemodialysis unit, appeared at times to "wear" the language, as an elite group member wears a badge of club membership. Such symbols not only unite and promote understanding between the members of a system, but add status and "specialness" to the actor's roles.

STRESS IN THE HEMODIALYSIS UNIT

Any social group encounters periodic system strains and the hemodialysis unit group, because of the nature of its composition (patients and caregivers) and the nature of its activities (related to life and death), might be considered subject to greater and more frequent stresses than many other systems.

In a study to evaluate the ward atmosphere of the hemodialysis unit, Kroemke and Nassar found that the primary problem reported was the patient's perception that "the environment was non-conducive to allow freedom of expression both verbally and non-verbally.[9] One suggestion to alleviate the situation was the formal scheduling of meetings between patients and staff where feelings of stress could be shared openly.[9] Lane and Hawkins, however, reported the most notable stresses in their hemodialysis unit to relate to reduction in professional staff and resulting overwork and unplanned overtime for many caregivers.[10] One method advocated for resolving this problem was the inclusion of the unit's liaison psychiatrist in the weekly nursing staff meetings in order to assist individuals to verbalize their feelings and frustrations: "Seeking assistance from an outside qualified source and working as a team helped us to identify problems, find solutions or ways to cope with seemingly insurmountable problems and thus resulted in the alleviation of stress in the hemodialysis unit."[10]

Although the research focus in the present study was on the maintenance hemodialysis experience, many caregivers admitted to experiencing strain in both acute and chronic dialysis units. In discussing the acute units, staff members frequently complained of being exhausted, overworked, and worn down from working in a crisis-oriented setting with extremely ill patients. Some caregivers suggested that in the acute units the pace was unsteady and thus stressful. It was reported that a unit might be very quiet for several days and thus tedious for the staff, and then suddenly there would be an influx of very ill "emergency" patients and the staff would be pushed to their limits to cope. One nurse stated,

> In an acute unit patient load varies so much. Some days it's real quiet, for example, 5 nurses and only 2 or 3 patients. That can get very boring. You can talk to the patients and teach them and so forth, but you can't do that for 8 hours. Then, at other times the unit gets hectic—it's crazy and you're just trying to fit in as many people as you can and that's frustrating because you know you can't do justice to everyone's care.

A hemodialysis head nurse working in a unit where both chronic and acute patients were dialyzed suggested that the dual focus of the unit sometimes resulted in stress, not only for the staff, but also for patients. She observed that dialysis patients always get upset if another patient has a problem on the machine and noted that this situation occurs more frequently with acute patients:

> If something happens to an acute patient, or if a chronic patient sees an acute patient come in very ill, you can just see that look of fear in their [chronic patients'] eyes, as if they were wondering if something like that would ever happen to them.

Even in dialysis units where the entire population is on chronic maintenance hemodialysis, patients watch each other and monitor each other's progress closely. Frequently, patients will ask about other patients if they appear ill, and some even watch each other's machines. One long-term patient stated, "We keep an eye out for one another while we're on the machine." A therapist commented,

> The patients watch each other to prevent things from happening. If something does happen to one of them you can sense very strongly from each of them that they are identifying with that person and thinking, "Well, maybe it will be me some day." And if a patient doesn't show up [for treatment] or is in the hospital, they [the patients] all know it before the staff. Their grapevine is phenomenal—it's very impressive. They have quite a network of communication among themselves.

A nurse in a chronic unit observed that she felt the noise level in the unit made both patients and staff members irritable at times and this resulted in stressful interactions. She said,

> Hemodialysis is noisy. It's all in one big room. There's conversation. There are a lot of machines. You have got water coming through pipes with pressure behind it. And the machines make all kinds of noises, with alarms and buzzers and all that. This setting would be enough to stress anybody out—patients or staff.

Several caregivers made the point that in their perception it was the long-term relationship with chronically ill patients that made the dialysis unit experience stressful. As Kroemeke and Nassar observe,

> The staff in a chronic dialysis unit relates to a relatively small number of patients for a long period of time. Consequently, the intensity of affect between staff and patients on a dialysis unit, combined with frequent crises, generates a high potential for anxiety, stress and maladaptive behavior.[11]

As a head nurse in the present study put it, "You can't help but be involved in the patients' lives, but you get tired and frustrated too, because they don't get better. It's like you'd like to see somebody just get well and go home. Some do have transplants, but often they're the patients you don't ever see again." Another caregiver added,

> One disruptive or difficult patient can tear up the place [unit], too. Like Mr. X., who always wants to come off the machine early. He's a perfect example. If he wants to come off, we have to take him off. There's no point in getting everybody else all upset because he sits there and he yells and he curses and he screams and you know you're getting upset. The staff are just about to crawl the walls, and the patients get all keyed up and upset.

Finally, for dialysis caregivers, perhaps the most notable complaint of stress in the dialysis unit revolved around overwork and lack of adequate staff to carry out the treatment procedure. A head nurse describes a typically "bad" day:

> It's been a busy day, people have called in sick, and you have all kinds of technical problems. A coil blows and you can't get the patient on [the machine] until late and somebody is sitting there waiting to be put on and other patients are saying, "I've got to get to work" or "My X-ray appointment is due," and you could just scream! So there's tension in the whole unit.

While patient stresses obviously also involve such caregiver stresses as noise, chronicity, and lack of adequate staffing, particular sources of patient tension and discontent were notably related to the following themes, as patient comments illustrate:

The new therapist. "I really don't like feeling like a guinea pig and have them practice on my fistula—one new guy almost blew it for sure. The turnover situation [staff turnover] is tough."

Time. "One thing I just hate is being late. I like to get off [the machine] at the time I'm supposed to get off. I hate to be held up. Like sometimes you are supposed to get off at 1:00 and somebody else gets sick and they have to take care of him, so you have to allow for that. That could happen to you, too. But sometimes it's just they start with somebody else first. They have a guy who wants to come off early, you know. But after four hours, I'm really ready to get out of here."

Caregiver attitudes and staff turnover. "Sometimes your blood pressure goes up instead of down when you come in here [to the unit]. There is a lot of staff turnover and sometimes you find nasty attitudes. Nurses' attitudes really bother me. Sometimes one of them will come in here and curse you out and then you go home and you're all upset. Then it takes about a month until you start to feel better about coming back."

Stresses for both caregivers and patients were found to be magnified following an occurrence of serious emergency situations or crisis in the dialysis unit. The two emergency situations that occur most frequently in the hemodialysis unit are described as "codes" and "air." ("Code" is used as a synonym for cardiac arrest; the visual presence of a large amount of "air" in a patient's dialysis access lines presents the potential for an embolus to enter the circulatory system.) Both of these occurrences are unsettling for both caregivers and patients. A code may be more or less stressful depending upon its resolution, i.e., survival or death; it always presents a frightening possibility for the patients observing. It was reported by one caregiver that "no one wants to sit in a chair after a patient has 'coded' in it." One female patient who had been experiencing some ambivalence toward kidney transplantation immediately agreed to the procedure after watching a good friend "code", even though the patient who had coded did survive. The "transplant convert" told the unit social worker, "If I ever had any question about wanting to get the hell out of here, that question went out of my mind today."

Air embolus, or the presence of air in the patient-to-machine access line, which may or may not result in a "code" situation, is also an ever-present potential enemy lurking quietly in the dialysis unit. Modern technology has insured that most artificial kidneys are now equipped with alarm systems to warn of the presence of air, but caregivers and many patients nevertheless maintain a careful watch for its occurrence during the treatment

procedure. Patients are also reported to watch each other for such emergency situations as those described above. One patient shift group's sense
of "community" or group solidarity, as often occurs in crisis situations, was
graphically described by a unit therapist:

> When we're doing "put on" [initiating hemodialysis], we've more than once
> had a coil leak or a rupture when our back was turned and another patient will yell
> "Look at that coil—do something!" You'd be surprised, they're very helpful. I've had
> a patient grab the syringe out of my hand and administer his own heparin so that I
> could get to another patient fast. I've had a patient take his own pulse and say, "I
> feel fine—you go take care of that [the other patient's problem]." They [the patients]
> realize that if it was them, that's what they'd want. And it's really great. That's what
> really makes it worthwhile. To watch these patients care about each other and really
> work with each other—especially when there's a crisis.

NORMS OF DIALYSIS PATIENT BEHAVIOR

While most hemodialysis caregivers stated that they did not like to
categorize patients or use labels such as "good" or "bad" patients, certain
norms for dialysis patient behavior were clearly evident in the dialysis units
studied. In general the "good" or well-adjusted patient was viewed as being
compliant with fluid, dietary, and medication regimens; coming to the unit
on time for scheduled treatments; not complaining excessively while on
the machine; and performing daily activities in a socially acceptable manner,
carrying out whatever family and work responsibilities his or her physical
condition would permit. In a study to identify attitudinal barriers in communication between dialysis patients and their caregivers, Roper, Raulston,
and Cramer present a profile of the "good patient" as determined through
combining the opinions of hemodialysis unit staff members. Some of the
"good patient" characteristics include "compliance with fluid, dietary, medical and drug recommendations; helps prepare the machine for treatment;
is on time for treatment sessions; is involved with other patients and their
activities outside of the unit; is in good humor; and is knowledgeable about
kidney disease and dialysis."[12] The maladjusted patient or "bad" patient, on
the other hand, reportedly is possessed of such characteristics as gross noncompliance with fluid, dietary and/or medication regimens; periodic skipping
of treatment sessions; excessive complaining while on the machine, frequent
attempts to cut treatment time short; or failing to maintain work and social
activities and responsibilities, which his or her physical condition would
permit.[20]

It was found in the present research that moderate noncompliance was
tolerated (though not condoned) by dialysis staff during holiday seasons—
particularly such holidays as Christmas and Easter. As one staff member noted,
"You can't expect the patients to live like this all year and never cheat,

especially on holidays. After all you have to have some enjoyment in life, and for dialysis patients food is a really big thing." Cautions were issued, however, and a story is told of one head nurse of a large dialysis center who called together all of his patients the day after New Year's day. He informed them that, according to their collective chemistries that morning, one more holiday binge like the recent one and they would all most probably be transferred to the "Great Dialysis Unit in the Sky." The statement was strong, but one might suspect, effective.

In discussing dietary and fluid restrictions and medication orders, most caregivers expressed the belief that the patient should generally adhere to the regimen as prescribed. One physician stated, "A good patient, especially a kidney patient, is probably a patient who is really compliant in all aspects, in terms of diet, medication, and fluids." Nurses expressed high levels of frustration with grossly noncompliant patients in statements such as the following: "A 'bad' patient to me is one who has an apparent total lack of concern for his or her health. I think it's one of the most frustrating kinds of things—the person that comes in [to dialysis] really overload-ed [with fluid] and you'd just like to shake him and say 'What the hell is wrong with you? You're wasting my time!' "; "Some patients who come in with fluid overload just do not comply no matter how much you talk, even if you tell them how bad off they will be. That's really frustrating because you just can't get through to them. Nobody's found the answer to noncompliance."

In terms of attendance at treatment sessions and maintaining activities of daily living as one's physical condition allowed, most dialysis caregivers strongly supported and urged self-responsibility. One shift supervisor de-scribed her attitude this way:

> I think a patient has to take some of the responsibility, of the initiative. I have a very firm policy on this shift: if you don't want to be on dialysis, don't come here and waste our time. If the patients come to me and they want off [the machine] and they're being unreasonable, I take them off. If I have a patient on dialysis who starts acting like a child, screaming, being obscene, I take them off the machine. I don't put up with it. I tell them, "I have 21 other patients in this room and they're not going to have to listen to you." It's a give and take responsibility.

An important source of stress in the hemodialysis unit relates to disparity in patient and caregiver attitudes toward norms of patient care behavior, which may be labeled expectational noncomplementarity. In situations of expectational noncomplementarity there are conflicting attitudes related to the actors' role expectations and behaviors. If a partnering occurs between a patient who expects to receive total care during the treatment procedure and a caregiver who expects to assist only with those things that the patient cannot do for himself or herself, frustration, anger, and even direct conflict may result. Such situations were found to vary according to openness to change in a hemodialysis unit environment and patterns of patient–caregiver communication that were supported.

NEGATIVE SANCTIONING

Related to the concept of norming for dialysis patient behavior is the phenomenon of "sanctioning." Sanctioning in the dialysis unit is an informal rather than a formal process and was articulated by many caregivers in terms of "teaching" or patient education. As one caregiver put it, "It may seem like I'm trying to punish them [the patients] if I turn off and don't give them a lot of attention if they're acting out, but I can't support that kind of childish behavior. I have to teach them to take some responsibility for their lives."

Negative sanctioning was reportedly carried out in numerous ways, from a caregiver being rather distant (interpersonally) with a patient, to direct lecturing and open anger about what was considered to be inappropriate or noncompliant behavior. Sometimes the sanction was severe, as the following hemodialysis caregiver's report illustrates:

> I think in terms of frustrations; the greatest one was when somebody came in [to the unit] fluid overloaded. God, you know, people used to think we [caregivers] were nasty. But, boy, we sat there and we waited until the very last minute before we put that patient on dialysis. Let him feel it. It's like teaching a child. And I think all of the staff were nervous about it, like, are you playing games? My God, what if they code? I don't think it's a punishment. I think it is almost a last-ditch attempt to say, "feel what it's like," and "this isn't a game we're playing, this is serious." Because if every time they [fluid-overloaded patients] came in, we whipped them right on the machine immediately, then they'd just say, "Oh, well, fine. I have no problems." They'll never realize how dangerous this thing is that they're doing. I think what we did is a desperation kind of move to say, "Look, if we can't teach you any other way, maybe this will get through."

While most negative sanctioning appeared to be initiated by the staff and directed toward patient behavior or what was considered to be misbehavior (norm violation), one head nurse noted that patients sometimes sanction or "punish" the staff by anger, acting out, or even noncompliance, if caregivers were perceived to have behaved inappropriately or uncaringly. She commented, "Sometimes the patients get angry with us. They think we don't care and they want to punish us, so they do things like going off their diets because they know we'll be upset when they come in overloaded. But sometimes they just need the attention and to know we care."

It appeared that for both patients and staff members the phenomenon of sanctioning was essentially linked to and expressive of the concept of caring.

ETHICAL DILEMMAS

Because of the life-threatening nature of end-stage renal disease, and the quality of life that the illness condition may effect, ethical issues involving treatment decisions and prognosis abound.

As noted earlier, in the decade of the 1960s, patient selection committees were designated as gatekeepers of the then "scarce resources," to allocate dialysis treatment for those who met the committee's identified physical and psychosocial criteria. One example of such a group was that of the pioneering artificial kidney center in Seattle, Washington. As the center could not physically or financially accommodate all of the candidates meeting medical criteria for treatment, an "Admissions and Policy Committee" was set up to formulate and apply "nonmedical criteria to select which medically qualified candidates would receive treatment."[13] However, with the advent of Public Law 92-603, the 1972 Social Security Act providing coverage of treatment costs for ESRD patients, the picture changed notably. Selection committees became obsolete and all renal failure patients became potential candidates for dialysis and/or transplantation. Now, as noted by Fox and Swazey, the Federal government became in effect "the new gatekeeper of renal transplantation and dialysis."[14]

With the decade of the 1970s and its more "abundant resources," new ethical issues were raised in regard to who should be dialyzed, related to quality of life. The question was asked, "Should all end-stage renal patients be dialyzed?" Fox comments that most physicians still report "that certain more biologically-based criteria not only are but should be used to discourage dialysis treatment for various categories of patients who it is medically assumed would not benefit from the therapy, and might even be harmed by it."[15] She points out, however, that extensive financial coverage is presently resulting in the fact that "virtually all criteria of negative selection are being abandoned" and suggests "that the medical profession appears to be rapidly moving toward the point where almost every patient with end-stage kidney disease will receive dialysis and/or transplantation."[15] This fact is often decried by dialysis caregivers, some of whom have very strong feelings against the choice of hemodialysis as a must for all end-stage renal failure patients. One staff member asserted, "It's terrible—it's getting like the dumping syndrome. They [the physicians] will put anybody on [dialysis]. The patient can be 90 years old, have a million other complications and really be waiting to die and they still end up on dialysis, once their kidneys fail." Another caregiver vented her frustration regarding, as she said, "when a family forces us to dialyze a patient who doesn't want to be dialyzed." She reported the following situation as an example:

> We had a patient not too long ago, an adorable little old lady, a typical *terminal* patient, and she would lay there for four hours hooked on to that machine and pray, "Take me Lord." And even a couple of the other patients said to me that they ques-

tioned God and his judgement in this case. "You know," they said, "why did He make Y. live when she didn't want to and take somebody who did want to [live]?"

While this may be an extreme example, it has some relevance not only to patient selection but also to patient (and family) choice—that is, the decision to live on dialysis or to accept the alternative of death.

Hamburger and Crosnier suggest that a patient's decision to refuse treatment in the face of death should not be considered suicide if the choice was made in a free and rational manner.[16] Holden, in discussing the ethical alternatives of dialysis or death, points out that patients facing life or death situations must be allowed to freely consider both atlernatives.[17] Schowalter et al have reported a case study of Karen, a 16-year-old dialysis patient, who after an unsuccessful transplant and with the approval of her family, decided to stop medical treatment "and let nature take its course." The authors conclude that there are situations when an adolescent patient's wish to die should be accepted by the physician, but suggest that further debate is needed on whether the decision for such "passive euthanasia" should rest solely with the physician or rather with a team of consulting specialists.[18]

In the present study no patients verbally expressed the desire to die or directly terminated treatment. One young male patient, however, indirectly chose the death alternative through gross noncompliance with his dietary and fluid regimen, and ultimately failed to return for scheduled treatment sessions during an extended period of time. His friend, another dialysis patient, described his decision this way:

He had some problems in his mind—he thought whenever he came out here to go on the machine, his friends would be running around with somebody else. So he started staying away from the treatments. First he just stayed away about one time a week—sometimes he would come and sometimes he just didn't show up. Then one time he just stayed away for about two weeks and they had to rush him to the hospital. He had about three heart attacks and then he just never recovered. He was a young guy but you know I think he just couldn't take it; he wanted to die.

Deaths related to hemodialysis regimen noncompliance are often viewed in the framework of suicide or an active decision to die, but Goldstein and Reznikoff suggest that "such behavior may be more fruitfully regarded as an attempt by the patient to reduce the anxiety resulting from the recognition of his tremendous responsibility in the treatment program."[19]

Not surprisingly, it is sometimes either the patient's family or even the professional caregivers who appear to have the most difficulty reconciling themselves to a decision to terminate treatment. Anger and Anger looked at the dialysis staff members' feelings of ambivalence toward such a choice and suggested that caregivers must learn to deal with "their own feelings and values before they can support a patient who chooses to refuse dialysis."[20] McKegney and Lange note that, while it is difficult for dialysis team members to accept the idea of a patient choosing death rather than living, "even more difficult for them is discussing such a possibility with the patient or his family."[21]

Many caregivers interviewed in the present study expressed their feelings of ambivalence, frustration, and even anger toward both a patient's conscious decision to terminate treatment as well as toward what one therapist labelled "suicidal noncompliance." A dialysis unit head nurse described the case of a grossly noncompliant patient presently in a hospital intensive care unit:

> I have one man in the hospital right now, in the I.C.U. This man has potential. He is able to walk around. He has got a young child. He has a girl friend that he lives with that he refers to as his wife. But he has just decided to give up. So he says, "If I have a taste for something, I eat it. If I want to drink something, I drink it." Right now he is dying. We hope we can save him, but we can't save him from himself unless he wants to live. It's very frustrating to see him wanting to die.

Another registered nurse working in a chronic dialysis unit shared the following case history:

> J. was a young patient, 26 years old. It was suicide because he was so noncompliant. I personally worked with J. Everybody worked with J. J. called every one of us every day. J. screamed for help. He came in for dialysis every day of the week during one week because he was in such bad shape with pulmonary edema—because he was eating the foods which he knew were killers. J. did die. And that has really affected the staff morale—his death. Because he got a lot of everybody's juice, you know. People worked and worked and worked and the staff would get really mad at him. And they would say, "Damn it, J., you are going to die." And other staff would try to reason with him or coerce him. Some would say, "I am too mad at you to talk to you today." But we really did care about him.

In discussing patient and staff reactions after J.'s death, the caregiver added,

> J. died two weeks ago and everybody has been kind of lost. The other patients really felt it. A lot of the patients said, "He was so young." And the nursing staff felt helpless with this guy because everybody knew J. was going to die. We knew he was going to kill himself. And yesterday Dr. Y. brought it up. He said, "J. didn't need to die, damn it."

In continuing to evaluate staff reactions to J.'s death versus another dialysis patient's conscious decision to die, however, a distinction was noted by the nurse respondent: "It is interesting the difference between those two patients. One made a clear-cut decision and could handle the responsibility, the other one [J.] was very manipulative." The former patient's choice was described in this way:

> We had a lady [M.] who said, "I've had it. [She had six kids.] I don't want dialysis

any more. I am giving it to God. God won't keep me alive." So the nursing staff said, "She doesn't come [for regularly scheduled treatments], so we are not doing her any good, really, if we only dialyze her once in a while." Then, her access failed and she refused to have access [revision]. I think the nursing staff felt as though they gave her the caring. They had provided her with the information [she needed] and that was all they could do. The physicians, on the other hand . . . [felt differently]. When M. was in a coma and her mother called and cried to Doctor Z., "Please get M.," Dr. Z. said, "Get her in an ambulance, and get her over here so we can try to do something." That was wrong. The nurses were upset by that. You know we were screaming at the doctor, "God, that's so wrong." The lady had made her choice!

The same nurse respondent commented finally,

We have to present the patients with their options. One of those options is death. It is no treatment but it is a real option. Dialysis is an extraordinary means. It is so costly. The patient has to be committed to it, as well as the family. If a patient particularly is not committed to it, he is going to die anyway.

One other factor related to the issue of treatment refusal or treatment termination that was mentioned by several caregivers was the question of the patient's ability to choose or decide on their future because of their physical and/or mental condition at the time. A staff nurse described her feelings this way:

We had a woman, she was in such pain and she elected not to be dialyzed. It was really difficult for the staff. It's difficult because of the fact that you know that the patient's going to die but you're not sure that patients are able to make the right decision because of the uremia, and especially when they're in severe pain. One they're dialyzed they usually feel so much better that it's hard to know. A lot of times it's just the stage that they're going through like denial in the stages of dying.

A hemodialysis unit physician commented,

Very few patients refuse [treatment]. Some patients that refuse eventually change their minds. Of course we offer it [dialysis] to them especially if we feel that they will benefit from it in terms of rehabilitation, but if they have refused in spite of our guidance, then it's their decision. We have to consider that the decision comes from them alone. But the problem is that it [the decision] might be due to the uremia; their thinking might be impaired. That's a concern. But, if after persistent explanation with the family members, we cannot convince them, then it is out of our hands from there.

While it appeared from caregivers' responses that any patient death was painful to deal with, those related to more tenuous decisions manifested by noncompliance with the treatment regimen seemed the most difficult. A direct conscious decision to terminate treatment was supported and perceived as a rational choice; a patient situation of gross noncompliance was disapproved and assessed as irresponsibility and lack of acceptance.

The environment of the hemodialysis unit constitutes a distinct sociomedical system containing within its boundaries an ever-changing panorama of life and its incessant struggle for continuance. Here strategic and aggressive war is waged against an illness that continually threatens survival. Inherent in the war itself is the quietly lurking potential for destruction. Patients and caregivers, both independently and in concert, wrestle with the issues of pain, suffering and quality of life, in attempts to find some meaningful undertones in the frightening experience at hand. While motives are not always understood and conflicts not totally resolved, the give and take of the actors, involved in reciprocally caring and understanding relationships, make up the complex yet sustaining social life of the unit. Patients may find in this place a measure of support and comfort as well as protection against the biological enemy with whom they vie daily. Caregivers may find in the setting an experience of courage and transcendence, which they discover in the face of overwhelming physical deficit. A day may come when hemodialysis units are no longer needed. Until that time, however, both patient and caregiver must continue to experience the life-controlling uncertainty of this complex sociomedical milieu.

REFERENCES

1. Loomis CP: Social Systems, Essays on their Persistence and Change. Princeton, N.J., D. Van Nostrand, 1960, p 4
2. Homans GC: The Human Group. New York, Harcourt, Brace and World, 1950, pp 88–110
3. Living With End-Stage Renal Disease. Bureau of Quality Assurance, H H S, Public Health Service, Health Services Administration, Washington, D.C., 1972, p 18
4. Hester C, Hines K, Liberson M: Music therapy in a dialysis unit. J. AANNT 5:97–100, 1978, p 97
5. Gutch CF, Stoner MH: Review of Hemodialysis for Nurses and Dialysis Personnel. St. Louis, C.V. Mosby, 1975, p 196
6. Rhodes J: Promoting patient independence. Nephrology Nurse 79:37–39, 1979, p 39
7. Brundage DJ: Nursing Management of Renal Problems. St. Louis, C.V. Mosby, 1976, p 100
8. Exupery A: The Little Prince. New York, Harcourt, Brace and World, 1971, p 84
9. Kroemeke GT, Nassar T: An evaluation of ward atmosphere in hemodialysis units. J AANNT 7:282-284, 1980, p 284
10. Lane CA, Hawkins A: Managing stress in a hemodialysis unit. J AANNT 8:36–37, 1981, p 37
11. Kroemeke GT, Nassar T: An evaluation of ward atmosphere in hemodialysis units. J AANNT 7:282–284, 1980, p 282

12. Roper E. Raulston A, Cramer D: Attitudinal barriers in dialysis communications. J AANNT 4:179–197, 1977, p 194
13. Fox RC, Swazey JP: The Courage to Fail. Chicago, The University of Chicago Press, 1978, p 208
14. Fox RC, Swazey JP: The Courage to Fail. Chicago, The University of Chicago Press, 1978, p 345
15. Fox RC: Essays in Medical Sociology. New York, John Wiley and Sons, 1979, p 139
16. Hamburger J, Crosnier J: Moral and ethical problems in transplantation, in Rapaport FT, Dausset J (Eds): Human Transplantation. New York, Grune and Stratton, 1968
17. Holden MO: Dialysis or death: The ethical alternatives. Health Soc Work 5:18–21, 1980
18. Schowalter JE, Ferhold JB, Mann NM: The adolescent patient's decision to die. Pediatrics 51, 1:97-103, 1973, p 102
19. Goldstein AM, Reznikoff M: Suicide in chronic hemodialysis patients from an external locus of control framework. Amer J Psychiat 127, 9:124-127, 1971, p 124
20. Anger D, Anger DW: Dialysis ambivalence: A matter of life and death. Am J Nurs 76, 2:276-277, 1976, p 276
21. McKegney FP, Lange P: The decision to no longer live on chronic hemodialysis. Am J. Psychiat 128, 3:267-274, 1971, p 273

──── 6 ────

Early Adaptation to Hemodialysis

Often the test of courage is not to die but to live.
Vittorio Alfieri
Oreste

Perhaps the most difficult test of courage for an end-stage renal patient is consciously to choose to pursue the continuance of life on maintenance hemodialysis. At first, the "artificial kidney" may appear as a savior alleviating the fearful symptoms of terminal uremia; however, as the physical condition begins to improve and one achieves a more balanced perspective on the future, the thought of total life dependency upon a machine may become overwhelming. The early hemodialysis patient sees in somewhat "hazy" relief a future in which previous family or career goals may have to be abandoned, life activities modified, and relationships are endangered. The prospects are not reassuring; the outcomes uncertain; the path lonely, but a choice must be made. One must accept or reject the fearful yet sustaining therapy made possible by advanced medical technology. The decision is awesome and only the courageous give assent.

STRESSES OF EARLY ADAPTATION

The Case of Chronic Illness

It has been repeatedly noted in the literature that illness forces some degree of disruption in ordinary patterns of interaction, and the normal fulfillment of roles and responsibilities may be impossible. "Job, home and community, the major locations of social roles, are all affected in one way or another by illness."[1] Therefore, it is concluded that as "a social and psychological event, illness is rarely uneventful, usually stressful and occasionally devastating."[1]

In a theoretic discussion of chronic illness, Shontz et al suggest that the position of the person with a chronic physical illness is as follows: "the individual is blocked by his disability from engaging in behaviors [energy exchanges] which are dependent upon the affected portions of the body. The illness is an active threat to the person to the extent that he attempts to overcome it directly and sacrifices his higher level developmemt to this end."[2]

Rodney Coe maintains that, "while the social–psychological effects of acute illness are usually of short duration, chronic illnesses are long-term conditions which may require new adjustments on the part of the disabled person and his family."[3] In equating chronic illness with disability, Coe points out that the chronically ill person tends to fall into a devalued social status and to encounter prejudice and discrimination, sometimes even from family and friends. He comments, "The acceptance of his devalued social role and an appraisal of himself as inferior are a reflection of the attitudes of persons with whom the disabled comes into contact."[4] In many cases of chronic illness the family has primary influence attitudinally, at least in those cases where involvement with family is prolonged and frequent. In such cases, family roles may be seriously disrupted and the chronically ill person may experience great stress in social–psychological adjustment processes.

In regard to the plight of the chronic hemodialysis patient, many unique sources of social and social–psychological stress may be identified. Often major adjustments in attitude and lifestyle must be initiated with the dialysis regimen, and interpersonal relationships between the patient and family and friends may be notably altered. Thus, the hemodialysis patient is often faced with a considerable degree of social isolation. Milton Viederman pointed out that one contributing factor to this isolation is the fact that the patients' survival is contingent upon their being attached to a machine several times a week, a machine from which they may never stray too far."[5] Viederman adds that "the machine itself almost seems to have a life of its own," and it has been found that patients often enter into a love–hate relationship with this "tyrant" that symbolizes the limits of their freedom.

Rajapaksa has identified a group of 7 stressors for the maintenance dialysis patient that include the illness itself, the machine, dependency upon the machine and helpers, dietary restrictions, sexual dysfunction, work and financial pressures, and the stress of waiting for a kidney.[6] Although several of these stressors have been identified and discussed in earlier chapters, the anxiety related to each is generally more severe during the early stages of patient adaptation to ESRD, when physical resources are frequently depleted.

During early adaptation to ESRD, patients must cope not only with their reduced physical strength brought on by uremia, but also with many fears, perhaps the greatest of which is that related to their "unknown" future on maintenance hemodialysis. Beard has commented that renal failure patients often fear an early death, but adds that those same patients "express their fears that even if they live, their lives may not be acceptable."[7] Levy has reported numerous specific psychological problems for the maintenance dialysis patient including "depression, ... high suicide rate, ... anxiety, sexual dysfunction, problems connected with difficulties in rehabilitation, the problem of 'uncooperative' patients and psychosis."[8]

Another situation that the dialysis patient frequently encounters early on in the adaptation process is insomnia, as pointed out by Friedman et al: "Nocturnal insomnia, coupled with daytime somnolence, fatigues the patient who then withdraws from many activities to conserve his ebbing energy."[9]

Finally, fledgling dialysis patients must deal with their treatment—necessitated dependence upon a totally new group of caregivers, the hemodialysis unit staff. As discussed in Chapter 4, dialysis unit staff members have different attitudes toward the hemodialysis procedure, the regimen, expected patient behavior, and quality of life for the dialysis patient. These attitudes may be either a help or a hindrance to patients struggling to cope with the early phase of adapting to life on dialysis. Kaplan De-Nour and Czaczkes, in studying professional team opinions among hemodialysis staff members, found that while some caregivers were very positive about the treatment regimen and considered their patients well-adjusted and rehabilitated, other staff personnel were of the opinion that "life on dialysis is a horrible thing,"[10] and that their patients were miserable.

In the present study, a considerable number of patient respondents were interviewed during the relatively early phase of their adaptation to hemodialysis. Of the original 126 study subjects, 34 (27 percent) had initiated dialysis during the previous year, and 27 (21.4 percent) had been on the machine between one and two years. Therefore, patient comments related to the stress of beginning the hemodialysis treatment regimen were picked up during the course of initial interviewing. Notably, there were patient complaints of "constant fatigue" and "exhaustion" following a treatment session. (It should be pointed out that as the study purpose was to examine social and psychosocial functioning among hemodialysis patients, subject selection criteria excluded any patients with serious medical or psychiatric complications, other than ESRD). Several "new" patients complained of severe nausea and leg cramps during the treatment procedure, and one patient experienced episodes of hypotension during each treatment session. A middle-aged female patient expressed her fear of the dialysis treatment session itself, commenting, "I start getting nervous early in the morning every time I have to come in here. It's scary to see your blood in that machine and sometimes I get sick and I never know what will happen. This is a bad disease." Another new initiate commented, "I've been coming [for dialysis treatments] for about six months now and it still takes everything I've got to make myself come in here. It's good [the machine]. It saved my life, but it takes a whole lot out of you, too. Sometimes you go home and you're just washed out. You have to lie down for about two hours." Another female patient remarked, "At first, on dialysis you're so scared you can't even talk. That's different for every patient, I guess, but that's how it was with me." A relatively large number of patients focused on "the machine" as their most immediate perceived stressor during their period of early adaptation. One male patient suggested that although he had some concerns about having a kidney transplant, he would gladly accept one to free him from the hemodialysis treatment. He stated, 'I'll go along with it [kidney transplant]— anything to get off the machine. I get itching and hot from the machine and feel terrible. My oldest son was going to give me a kidney, but he was sick too much." Another patient reported that, although he had previously been negative in his attitude toward having a kidney transplant, he now wanted one: "What changed my mind was dialysis and how you feel awful the day after." A 31-year-old married male patient linked his continued sexual problems to the machine, noting, "Those blood pressure pills made me impotent. After I got off my blood pressure pill things got some better, but I still have problems. I have a lot to contend with as far as the machine, the treatment, and I often just don't feel like it because of that." Another male patient, married, aged 59 years, also discussed his sexual problems in relation to the treatment procedure, saying, "The machine does you in. There is none [sexual potency]. It's a pain. A woman can say it's not a problem, but a man can't say that because he's the one who has to produce. I wouldn't even think about touching a woman." When this same patient was asked if anything positive or good had come out of his illness experience, he replied, "No, I would like to die."

Several dialysis patient respondents complained about the length of time scheduled on the machine and said they would like the treatment time shortened; and a number reported that medications, especially antacids, were difficult to take and made them nauseated. While some patients were able to maintain their usual work activities and fulfill expected role responsibilities, most of this early adjustment group commented that it took all of their energy to keep up. One male patient in a professional occupation noted, "I can keep up my regular 40-hour work week, but my social life is 'nil.' By the time the weekend comes, I am ready to collapse and I just rest up for Monday morning." Another patient with a professional career admitted, "I do get frustrated about the time I have to spend in here [dialysis unit]. I could be doing a lot of other things and sometimes I get depressed about this." A few patients expressed concern about finances, the futures of their children, and their own prognosis. One patient, however, pointed out that his attitude toward life on dialysis was influenced by the fact that he had freely chosen to become a hemodialysis patient. He put it this way: "I'm not overly bitter about dialysis, due to the fact that ethically you can do what you want. I see this as a burden, but not as a catastrophe."

Almost all of the patients in the early adaptation phase of the dialysis regimen complained about the stress related to adherence to a restricted diet and fluid intake.

EARLY REGIMEN COMPLIANCE*

In the present study, early regimen compliance for the hemodialysis patient was measured quantitatively by means of a 7-item scale based on the patient's self-report of behavior for such factors as attendance at scheduled treatments, adherence to dietary and fluid restrictions, and taking of prescribed medications. It was discovered that early regimen compliance among patients, including dietary and fluid regimen adherence, was significantly linked to the support and expectations of significant others. Correlational analyses of T1 interview data revealed statistically significant relationships between the patient's perceived support from family members and caregivers and actual patient behavior in regard to compliance with the therapeutic regimen. The relationship between perceived support by significant others and compliance behavior was also analyzed, controlling for the educational level of the dialysis patient. Statistical analysis presented a pattern that demonstrated that the perceived expectations of primary group members, mainly family, were more highly associated with compliance behavior of upper educational-level respondents; but the expectations of members of the secondary group system, i.e., the caregivers, were more strongly correlated with the behavior of the lower educational-level patients. Although it has been observed that strong primary group support systems are often present within the lowere SES[†] extended family system, it would appear from the above findings that for this sample group such support did not effectively extend to sick-role behavior as operationalized in terms of compliance with a regimen. A possible explanation is that the poor frequently

---

*This section is abstracted notably from the author's doctoral dissertation (O'Brien ME: Hemodialysis and effective social environment: Some social and social psychological correlates of the treatment for chronic renal failure. The Catholic University of America, Washington, D.C., 1976, pp. 157–161).

[†]In the present study many of the dialysis patients interviewed initially (T1) were unemployed and income was primarily dependent upon federal financing. Thus, of the three indicators of S.E.S., education, occupation and income, education was considered the most stable and employed as the primary analytic indicator of patient socioeconomic status.

have more immediate problems to deal with than planning for the specific details of a regimen to maintain health. Hurley, in discussing the health crisis of the poor, asserts that priorities other than medical care of treatment come first. He states that, "The dilemma of the welfare recipient is shown by the fact that every cent of the allotment must be used for immediate needs," and he notes that in the final accounting, "not enough money is left for a single visit to the doctor or for scarcely any other health measure."[11] R.J. Haggerty is of the view that one reason the urban poor place much less importance on medical treatment than middle-class persons is the educational and attitudinal barriers on the part of the less-advantaged patient.[12]

Thus, for the family of the low SES chronic renal patient, simply having enough food to survive may be more important than maintaining the special low sodium, low potassium diet prescribed for the physical well-being of the dialysis patient. Often, also, members of the low SES family or primary group may not have sufficient accurate knowledge relating to a medical regimen and are therefore unable to give necessary support, by way of expectations, to the ill family member. Kosa et al develop the theory that people assess medical problems with the aid of the knowledge that they have on matters of health in general, and add that this knowledge is unequally distributed among social classes: "it is a part of those privileges which mark the differences between the poor and the non-poor."[13]

In descriptively evaluating the initial group of 126 study respondents in regard to compliance behavior, it was found that *no* statistically significant differences existed on compliance scale responses related to the following: length of time on dialysis, type of household (patient living alone, with adults, with children only, with adults and children), income, occupation, education, marital status, race, age, or sex. Mean scale scores did reveal certain sociodemographic trends, however. When compliance behavior data were analyzed according to age of the dialysis patient, lowest mean scores were found among the younger patients, with the highest compliance response reported in the 60–69 year age category. There was only a slight distinction between males and females, with males slightly more compliant; as well as between blacks and whites, with blacks somewhat less compliant that their white counterparts. Married dialysis patients reported the highest compliance response, while averages for other marital status categories were divergent.

In a review of the literature dealing with compliance, Marston reports that "the results of most investigations have led to the conclusion that age is probably not significantly related to compliance."[14] Several studies do suggest, however, that younger patients are less likely to follow treatment regimens than older patients.[15] A number of authors reported no significant relation between sex and compliance[18]; the same was generally true for marital status, with the exception of Morrow and Rabin, who found married persons more likely to comply than separated or divorced patients.[17]

In looking at socioeconomic status, assessed in terms of the 3 major components chosen for this study, one finds that, regarding education, the highest compliance scores were reported for those persons having approximately two years of college and, regarding occupation, the highest mean score was recorded for professionals. Complaince scores related to income categories, however, do not follow the pattern—dialysis patients with the highest mean response scores fell at either end of the continuum, i.e., $2000–$3000 or $25,000 and above.

Previous studies have not demonstrated that socioeconomic status is importantly related to compliance. A few researchers do nonetheless report that higher levels of education were associated with greater degrees of compliance to treatment. Milton Davis, in a review of compliance literature, notes that is a significant percentage of the studies considered, the authors discussed educational level of the patient and "all agreed that higher education was related to compliance."[20]

In evaluating variations in compliance data in relation to type of household and duration of treatment, it was found that persons who live alone had the lowest mean response score for compliance with the treatment regimen; those living with other adults and children had the highest score. Scale mean responses in regard to length of time on dialysis appeared to be randomly arranged, with the lowest score appearing in the "over four years" category.

In a study of hemodialysis patients dealing with compliance and rehabilitation, it was found that, irrespective of all other factors, patients with several children "have obtained better than average rehabilitation."[21] As to treatment history, Marston suggests that, "there are some indicators that length of time under treatment influences compliance."[22] Several investigators report that compliance decreases over time.[23-24]

EVALUATION BY DIALYSIS CENTER PERSONNEL

As is the case in much contemporary behavioral research, the findings for the present study were based on self-reported data obtained through patient interviews. In order to evaluate data on compliance with the treatment regimen, however, it was felt that an independent judgment would be especially appropriate. Not surprisingly, patients often tend to overestimate their degree of compliance in an attempt to portray themselves in the role of the "good patient."

As dialysis center staff members are generally quite busy, the investigator deemed it inappropriate to subject them to the particulars of the patient compliance scale. Therefore, a structured interview item dealing with the patient's overall compliance to the regimen was utilized for the staff interviews.[25] A dialysis center staff member assigned to the respondent in question was designated as "judge" and his comments were confirmed by recent weight records for each patient.

Evaluation of patient compliance data derived from the caregivers showed that of the 126 patients originally interviewed, judges reported 11 patients (8.7 percent) as never complying with the regimen, 54 patients (42.9 percent) as sometimes following the regimen, 47 patients (37.3 percent) as usually complying, and 14 (11.1 percent of the patients were judged as always adhering to the prescriptions of the treatment regimen. From these data one observes that a relatively small number of patients, 35 (19.8 percent) fell into the extreme categories of always complying or never complying, with the great majority (101 or 80.2 percent of the total sample) falling in the middle compliance categories.

Judgments of the dialysis center personnel differed somewhat from patients' self-reports, as can be shown by analyzing one item with the compliance scale—patients' self report re overall participation in the dialysis regimen. Patient responses were as follows: 3 patients (2.4 percent) reported "never" complying, 36 patients (28.6 percent) "sometimes," 64 patients (50.8 percent) "usually," and 23 patients (18.2 percent) reported "always" complying with the regimen. When comparing these responses with those of the dialysis center "judges," patient self-reported data tend to show somewhat higher rates of compliance with the hemodialysis regimen.

ACCOMMODATION TO A HEMODIALYSIS LIFESTYLE

How chronic dialysis patients will cope over the long run, what lifestyle and what attitudes and behavior they will adopt, is often largely influenced by early attitudes and behavior related to acceptance or rejection of one's condition. Admittedly, during the first few days or even weeks of dialysis, patients may still be experiencing symptoms of fatigue, nausea, or depression related to the uremic condition. As these symptoms gradually are alleviated, however, patients begin to consider more carefully their life on dialysis. They must now make a decision about that life—a conscious decision to live, to survive, if the prescribed hemodialysis treatment is to have meaning.

Pritchard has asserted that the chronic dialysis patient may view his illness in one of two ways: "On the one hand, he may see it as an enemy which is attacking him, or on the other, as a challenge or problem which he must tackle with whatever resources he has," and he notes that each attitude has consequences for how the illness will be dealt with.[26] Mlott pointed out that not only must the maintenance dialysis patient deal with illness—fatigue, lack of energy—but also with a number of problems that involve family, work, and social activities. He notes that the dialysis patient may have to withdraw from certain social groups and might also be faced with "failure of contemplated plans or ventures such as an inability to purchase a home which was anticipated, as well as cancellation of extended vacations, which invariably include spouse or family members."[27]

For patients involved in the early phase of adaptation, accommodation to life on dialysis may prove more difficult, as they have not yet learned to initiate particular coping mechanisms that may be adopted later on. Gentry and Davis have reported that patients who had experienced dialysis over a longer period "were characterized by a greater tendency toward responding in a socially desirable fashion and being less sensitized to the stressful aspects of their current life situation" than were patients who were newer to the dialysis treatment regimen.[28]

Several authors have identified and described phases or stages of adaptation to maintenance hemodialysis, the most notable perhaps being Abram[29] and Reichsman and Levy.[30] Abram presented a typology beginning with early patient adaptation, noting the following phases: "Phase I: The Uremic Syndrome"—patients in the near-terminal stages of uremia; "Phase II: The Shift to Physiological Equilibrium"—(dialysis) patients about the third week who are beginning to experience a return to a more positive physiologic and psychological state; "Phase III: Convalescence—rturn to the living"—third week to third month, when the patients' condition begins to stabilize and they must assess their life situation; and "Phase IV: The Struggle for Normalcy—the problem of living rather than dying"—(third to twelfth month) when patients have adjusted to the dialysis routine and must focus on the work of living. Reichsman and Levy, in a 4-year study of 25 chronic patients, distinguished 3 stages of adaptation to maintenance hemodialysis, which they labeled "(1) the 'honeymoon' period, (2) the period of disenchantment and discouragement, and (3) the period of long-term adaptation."[30] The "honeymoon" phase, which the authors describe as a time of psychological and physical improvement, marked with hope and happiness, was initiated "one to three weeks" after the patient had begun dialysis; the "period of disenchantment and discouragement," when patients began to experience depression and sadness, occurred following the "honeymoon" and lasted "from about three to twelve months." The "period of long-term adaptation" was considered to be that time when patients accepted their condition realistically. The authors note that patients' arrival at this latter stage was varied and gradual.

In the initial phase of the present research (T1, N = 126) all patient subjects had passed out of the "uremia" phase and most were also over the initial euphoria as described in the "honeymoon," the majority having initiated dialysis at least 5 to 6 months prior to interview. A number of study respondents, however could be described as in a period of disenchantment

with the dialysis regimen; some were initiating the phase of "long-term adaptation" or "the struggle for normalcy." one patient who notably exemplified
the stage of disenchantment was a middle-aged male patient, who was married, had been on the machine for approximately 8 months, and had a pale,
almost cachectic appearance. This patient voiced many complaints during
the treatment procedure, and related such symptoms as leg cramps, nausea,
and headaches. He was admittedly and, as described by dialysis unit personnel, grossly noncompliant at times. He had been known to come in to
the unit 5–6 kilos over his target weight. He appeared to have fairly strong
family support, but did not respond well to attempts at communication initiated by dialysis center personnel. This patient appeared not to have accepted his illness condition and the treatment regimen as reflected both in
the behavior described and his own comments. He stated, "Sometimes I
really just feel like giving up. I feel bad all the time. I'm not getting any
better and I just wonder if it's worth it." The patient expired after approximately 14 months on dialysis, probably never having entered the period
of long-term adaptation.

Another male patient, married, 52 years of age, appeared quite depressed
and commented negatively about his illness condition and previous year of
hemodialysis. He noted his perception of the inadequacy of the kidney
transplant procedure and complained about the lack of attention he received
from the "doctors." In regard to family support, this respondent reported
that he "didn't have too much." He stated, "My sister mentioned giving me
a kidney a year ago, but hasn't said anything since. I feel like I can't ask. I
just have to wait and see if it's mentioned again." The patient was also depressed about his self-reported sexual impotence. When asked if he had any
problem with sexual functioning, he responded, "Why should I comment.
You know darn well there isn't any [sexual potency] and anyone who says
there is, is a liar. I'm totally impotent." Then he added, "how can I have sex
if my back is killing me." The patient, finally, commented on his overall
condition as follows: "I'm depressed. Why did this happen to me? I'm not
a religious man, but I always used to help people. I can't help but ask, 'Why
did this happen?' " The patient expired post-transplant, approximately 9
months following the interview.

A third patient who also presented the classic picture of "disenchantment and discouragement" was a 22-year-old male patient who was single,
reportedly lived alone, and had been on dialysis for a little over 2 years.
This patient admittedly was noncompliant with the dietary and fluid regimen
prescribed. He commented, "I eat what I want." He was also assessed as
highly alienated according to the standardized Dean Alienation Scale utilized
in the interview. He has not working at any regular job and rated his social
life as inactive. The patient noted that he got along "very poorly" with members of his family, admitted to no church or club activities, and identified

no religious affiliation or orientation whatsoever. He reported that his recreation time was more often spent alone, that he was sexually impotent, that he did not want a kidney transplant, and commented of himself, "I'm bitter toward the world." Several months following the interview, he intiated a pattern of missing treatment sessions, not complying with dietary and medication prescriptions, and after several serious episodes of pulmonary edema, was finally brought to an emergency room where he expired. This patient appeared to have experienced an extended period of discouragement from which he was not able to recover. Also, he did not appear to have any significant family support to help him to cope with his condition and did not communicate well with dialysis center personnel.

As patients begin to progress beyond the early adaptation phases of euphoria and discouragement, necessary accommodations in thinking and behaving were adopted. A 59-year-old business executive commented on the difficulty of having to "slow down" his lifestyle: "At first it was an overwhelming sense of frustration for a successful man, but then you begin to accept." This patient reported that he felt he had successfully learned to cope with the dialysis regimen after a little more than a year, and noted that he now believed he treated his employees more humanely, "more as persons" because of his own suffering and illness experience. A younger male patient commented, "I have my routine. I try to keep myself active. I feel better both mentally and physically. You don't get depressed if you stay involved. If you get on a lot of fluid you get tired. That's why I try to stay active."

Another young male with a professional career, who had been on dialysis approximately 2 years, noted that while he was coping well his illness and treatment regimen, certain restrictions had to be accepted. He observed,

> This affects the type of job I can hold, the type of social life I can lead. Your ability to travel is restricted. Your food is restricted. Food can become an obsession. Life in general is a lot more difficult. You often become tired and weak. People really don't realize how you feel in between dialysis [treatments]—they don't have any appreciation of this.

A 63-year-old male patient who was married, had been on the machine over 3 years, and was reportedly very compliant in regard to dietary, fluid, and medication regimens, complained primarily about the difficulty of dependency and lack of ability to travel freely. He observed that as a professional person he had "done a great deal of traveling," and noted, "I don't have the mobility I had before." This patient commented that he felt he was "limited in time and energy" but had adjusted his present lifestyle to the hemodialysis regimen. He also reported on his acceptance of and adjustment to his illness condition:

> Over and beyond social beliefs, there's a personal commitment to carry on; a

responsibility to family and others around you. It doesn't depend only on yourself, but acceptance is a function of your responsibility to others.

Finally, a 52-year-old female patient who had been on the machine approximately 6 years at the time of interview commented on her long-term adaptation as follows:

It really took me about a year and a half to adjust to dialysis. I used to get real sick on the machine. I would be nauseated and vomit and sometimes my blood pressure would go way down and I'd get real depressed. But you have to live one day at a time. At first I used to just think about coming to dialysis and nothing else all day, even the day before. Now I think about all the things to do at home and only think about it when it's just time to come.

This patient demonstrated a marked improvement in attitude and behavior over the years and has recently initiated "self-care," which she commented on:

I'm on "self-care" now and I love it. I have so much more freedom this way. I like to stick myself and watch my own machine. "Sticking" is important and used to be a problem for me. If they don't stick you right, it hurts all during the treatment, or if they get it against the [arterial] wall you don't get a good flow, but now I can do it myself.

This patient reported that she was relatively content with her life on dialysis now and her only problem was continued insomnia. She commented, "It's kind of hard for me to get up in the morning. I like to sleep in. At night I toss and turn. I haven't slept well since I've been on dialysis. I guess I worry about some things."

Overall it was found that study patients' attitudes and behaviors fluctuated to some degree over time in relation to the presence or absence of more serious physical or psychosocial problems such as hospitalization and surgery or the loss of a loved one. Once a basic attitude and pattern of behavior was established in regard to ESRD and dialysis, however, deviations from the norm were generally temporary and quickly resolved once the initiating problem was alleviated.

"SICKNESS–WELLNESS" SELF-PERCEPTION CONTINUUM

In the present research it was observed that chronic dialysis patients generally adopted 1 of 3 modes of self-perception regarding ESRD. First, there is the self-perceived mode of "sickness," in which the patient's primary focus and interest are upon bodily response—survival. In this mode patients perceive and report themselves as "sick" and as fighting to alleviate their illness condition; a second mode relates to the patients' perception of having a "chronic illness," in which focus is placed upon learning about and adhering to the regimen and the rearranging or reorganizing of one's lifestyle to cope with the disability. In this mode, patients describe themselves chronically ill, but report a more positive outlook for the future. Third, patients may adopt a mode of self-perceived "wellness" in which the greater part of their interest and energy are zeroed in on life activities and goals, and/or the family, and ESRD and dialysis are secondary. These three modes may be perceived as comprising a continuum upon which patient attitudes may be located:

Self-perceived	Self-perceived	Self-perceived
Sickness Mode	Chronic Illness Mode	Wellness Mode

<-->

It was found that some study patients moved unidirectionally from the phase of self-perceived "sickness" to a perception of "chronic condition" and even on to the self-perceived "wellness" mode within 1–2 years of initiating the hemodialysis regimen. Certain patients, however, retained the self-perceived "sickness" attitude throughout the course of the research, which for some could mean up to 11 or 12 years following the diagnosis of ESRD and the start of dialysis. The comments of one long-term dialysis patient reflected the concept of the "sickness–wellness self-perception continuum" rather directly:

> I don't think "sick." Since I've been on the machine I've never really been "sick." You can be "sick" on dialysis or you can be "well" on dialysis. It's up to you. You gotta make up in your mind what you're gonna do. Nobody is gonna do it for you.

Another male patient described his early adaptation to dialysis and commented that he did not now consider himself to be "sick" as he was working as much as 16 hours a day running his business. He reported, "I didn't tell myself that I was sick. I never let my mind be conditioned that I was sick. How soon you adjust is the condition of your mind. If you feel negative it takes a lot away from your life." He added, "The patient has to think positive and be positive—that's the key to the word 'coping'."

Landsman reports that counseling sessions with over 100 patients with renal disease revealed a notable problem in self-image. She observed that patients "tend to find themselves adrift somewhere between the world of the sick and the world of the well."[31] Such a sense of vascillating between two worlds of sickness and wellness was also observed in certain study patients discussed in the present monograph. The "sickness–wellness" continuum concept is presented to identify a typology of basic self-perceptions revealed by the majority of study respondents. Obviously, patients moved back and forth along the "sickness–wellness perception continuum" in re-

Table 6-1.
Attitudinal and Behavioral Correlates of the
Hemodialysis Self-perception Modes

Self-perceived Sickness Mode	Self-perceived Chronic Illness Mode	Self-perceived Wellness Mode
dependence	acceptance	independence
anxiety	trust	control
anger	cooperation	integration
alienation	active involvement with regimen	social involvement
withdrawal	family/friendship involvement	work/career involvement
problems "on the machine"	few difficulties "on the machine"	self-care
"ritual" compliance*	"modified" compliance	"reasoned" noncompliance

*Compliance types are discussed in Chapter 7.

lation to physical and psychosocial stresses that occurred in their lives. There did not, however, appear to be any pattern of relationship between the basic modes of self-perception or movement along the continuum and sociodemographic variables such as age, sex, race, religion, marital status, or socioeconomic status.

Certain basic attitudinal and behavioral correlates of the hemodialysis patient self-perception modes were noted in study respondents. These are presented in Table 6-1.

A number of individual comments illustrate the rationale for patients' adopting particular self-perception modes. One male dialysis patient noted, "When I first started [dialysis] I thought you were *supposed* to be sick, supposed to walk around with a cane and rest all the time. I wasn't sick, I just needed something to eat. I didn't get enough to eat." A female patient observed,

> I used to be sick because I decided I was sick; so, I got nauseated on the machine, I used to vomit constantly, was always tired and all that. Then I decided to live. I know I have a chronic illness and have to take a treatment for it, but I'm not sick.

Another female patient commented, "Someone should tell the patients sooner [after starting dialysis] that 'you can do'—you're not going to be sick the rest of your life." A long-term male dialysis patient described his experience this way:

It took me about six months to adjust to this [ESRD and dialysis]. I couldn't believe I was sick. I had never been sick before. It all hit me one day and I really knew I was sick. But now I'm not sick. I don't see myself as somebody who's sick. I just have to take a treatment.

An older retired patient described himself as "well." He stated, "I'm feelin' fine. Yes, I'm a well man. I know what to eat and don't miss a treatment and I enjoy my life. I can drive my truck around and work in my garden and go where I want to. This hasn't been too big a problem for me."

One younger male patient who had been on dialysis over 3 years perceived his life quite differently. His self-report represents the "sickness" end of the continuum:

I'm a sick man. I don't do anything but come to this unit anymore. I can't eat. I can't work. I can't even take care of my family. So what else is there to say? When I come home after this machine I'm so worn out all I can do is go to sleep and then I have to rest up the whole next day until it's time to come back here again. So that's where my life is now. It's not a pretty picture, but what can I do? You can't help being sick.

As noted, study patients were observed to change self-perception modes and to engage in movement along the sickness–wellness continuum. A female patient described her movement this way:

At first I decided that I was sick—for about two years. Then I didn't what that any more. I decided to stop that foolishness. God brought it to me that I was killing myself. I've gotten my weight down. I stick to the diet. I still need people to help me, but I'm much better than I was. I just have a chronic illness.

A hemodialysis nurse validated the concept of patient self-perception modes in commenting, "There is a 'sick' dialysis patient and a 'sick-well' dialysis patient. I think it mostly has to do with how they see themselves." A family member commented on the importance of providing support for the patient in terms of a wellness self-perception. She stated, "You have to treat them like they're normal—like they're basically well people." She added,

If you're treating them like they're sick, then why not be totally sick? Instead, you should tell them they're capable of doing this and that, then they will do it up to the point where they feel they're not helpless; they're still very functional and then they have a tendency to go on for who knows how long. I think that's the most important attitude—facing the illness and still being content with yourself. It shouldn't destroy your worth. There's nothing wrong with a little sympathy, a shoulder to cry on, to know that someone is being tender, just don't drive it into the ground.

SECONDARY GAIN*

In discussing the concept of "secondary gain," Duff and Hollingshead explain how certain ill persons "use" their condition to achieve desired ends within the family while other patients are "used" as objects of control by well family members.[32] King comments that being dependent may satisfy an ill person's "childish" need for attention, and may promote dependency to such a degree that the patient encourages symptoms to persist beyond the expected course of the disease.[33] Mechanic asserts that it is possible for an ill person to achieve secondary gain not only for himself or herself but also for other family members, and cites as an example the case of a father who, through physical incapacity, has more time to spend at home with his children.[34]

In regard to the hemodialysis patient, Kaplan De-Nour and Czaczkes found that patients sometimes use their conditions to solve, to a degree, their dependency–independency problems. They suggest that dialysis patients "get or try to get 'secondary gain' from their conditions—more attention, social benefits from society, etc. This tendency, which is rather common in chronic patients, may cause or increase the abuse of the medical regimen," and they add that, "There are some patients for whom dialysis does not create problems, but for whom it solves predialysis conflicts. These patients do not want, usually, unconsciously, to get well, and thus have to face their problems alone, and this might lead to the abuse of medical regimen."[35] Reichsman and Levy conclude that, for most hemodialysis patients, becoming productive again presented a major conflict. They hold that, "the patients' pre-conscious or unconscious wish was to continue in the dependent role."[36]

Dansak asserted that secondary gain must be decreased in order to promote successful rehabilitation of the patient on long-term hemodialysis.[37] He cites Abram, whose discussion of the patient's convalescent phases includes that aspect of the dependency–independency conflict termed "secondary gain." For the patient to adapt successfully to dialysis, Abram states, "it is necessary to decrease the secondary gain of this illness and allow his healthy independence to become dominant."[38] Anger and Anger reinforce this statement in suggesting that the benefits referred to as "secondary gain" must be reduced in order to promote successful rehabilitation. They also point out that forfeiting secondary gain may involve a risk of reducing the patient's sense of security, and they assert that "unless sufficient motivation

---

*The discussion in this section is abstracted notably from the author's doctoral dissertation (O'Brien ME: Hemodialysis and effective social environment: Some social and social psychological correlates of the treatment for chronic renal failure. Unpublished doctoral dissertation, The Catholic University of America, Washington, DC, 1976, pp 37–39; 124–129)

can be aroused to counterbalance the secondary gain, the patient will continue to maintain the sick role and will retain the benefits of that role."[39]

Admittedly, secondary gain is a difficult concept to identify or evaluate when dealing with self-reported data. Even the patients themselves are frequently not aware of certain "satisfactions" that may result from having to cope with an illness condition and its consequent behavior modifications.

Even if the patients are conscious of some satisfying or positive benefits of the illness experience, they may be loathe to discuss them for fear of negative sanctioning. Because of the import of secondary gain, however, whether it be either positively or negatively associated with rehabilitation, an attempt was made to explore the concept in the present study. This was done by including several interview schedule items that might stimulate patient response in the area and perhaps provide some insight into any satisfactions that could accompany chronic renal failure and treatment.

In analyzing the data statistically, it was found that mean responses on a "secondary gain scale," analyzed according to age, showed highest scores for the 70–79 year category. Mean scores by sex and race were basically undifferentiated. For marital status categories, the highest mean score was found among persons who had never been married.

Mean responses for secondary gain evaluated by education showed the lowest score reported for the group having at least 2 years of college and the highest for the "grammar school only" category. The findings for occupation were similar, which lowest scores reported for professionals and highest scores for unskilled workers. Mean scores by income showed a random pattern of distribution.

Scale responses for type of household revealed little variability; patients who have been on dialysis 5 years or longer report the highest degree of secondary gain.

As well as quantitative data obtained through utilization of the "secondary gain scale," a notable amount of qualitative material was obtained in the form of study patients' responses to the interview questions dealing with secondary gain. The first question was: "Have any positive or satisfying results or 'good things' come out of your illness and dialysis experience, either for yourself or your family?" The patient was then asked to elaborate on the above with the request, "Could you tell me about this?" Many patients appeared to grasp the issues involved in secondary gain, as the following comments illustrate. One male study respondent, 64 years old, married, on dialysis over two years, reported, "My relationship with my family is much closer now—we have better cooperation. I spend more time at home." A female patient, 52 years old, widowed, on dialysis over 3 years, observed, "I think this has made me think about life and be grateful for life and grateful to God. It's made me find things to do for other people besides worry about myself." A male, 38 years old, divorced, on dialysis 2 years, noted, "I have more time now to be with my kids." Another male, 59 years old, married, on dialysis over 2 years, described it this way:

> At first it [illness] was an overwhelming sense of frustration for a successful man, but then you begin to accept, and I began to think and appreciate every day that I have extra to live. Kidney disease has changed my values and made me more considerate of others."

A 47-year-old male patient, who had been on dialysis a little less than one year, commented that one good result of his condition was that he was forced to take some time to think about his life and the consequences of his actions:

> This illness definitely made me think—get my mind together. I know all things happen for the good. It turned me around spiritually and mentally. Now I listen better. I try to be more patient and I have more to learn from others. I tend to listen to instructions better.

A female patient who was also relatively new on dialysis reported, "I have more free time now. I have become closer to my family. They expect more of me than before. I have become closer to my husband and children." As the previous respondent, many patients in discussing secondary gain focused on the impact of ESRD and dialysis on family interaction: "It's made me get much closer to my sister who is on dialysis. We help each other"; "I get to see a lot more of my family than before I was ill"; "It has made my family much closer—really brought us together."

Two patients commented on secondary gain in the context of recognition of their illness condition. One young male patient noted, "My disease was recognized and I'm getting help for it. My dad died from it and they didn't know why." A 34-year-old female patient observed, "I think it's been good facing the hereditary disease. Now my two sisters know they have it. My mother would never tell us about the disease, but this has brought it out into the open. My sisters, now that they know, they are being checked out." Finally, a number of patients focused upon the fact that their illness condition had made them think about and appreciate the value of their lives. A middle-aged male patient, who had been on the machine about 18 months, commented, "It has made me appreciate my life. It makes each day of life more important." A young male patient, on dialysis less than one year, described his attitude toward the illness experience:

> It has enlightened me as to just how fast I was really going. It made me re-evaluate my life. Now I can place my needs before my wants. It hasn't been so difficult in looking at the good advantages. This thing [illness] has made me think a lot about the way I used to live and put different values on things.

Overall, close to two-thirds of the dialysis patients in the original study group reportedly perceived some "good" either for themselves or their families as a result of treatment for ESRD. Some patients, however, did describe the illness experience very negatively and perceived nothing good associated with their condition. As one middle-aged male patient, who had been on dialysis approximately 2 years, observed, "I can't think of anything good about it, except maybe the fact that I've survived."

As so poignantly expressed by Alfieri (*Oreste*), "often the test of courage is not to die but to live." Striking is the courage of the early maintenance dialysis patients who consciously and completely choose to live amid the stressors associated with their illness. Among these are the frequent occurrences of pain and anxiety, coupled with a continuing uncertainty of the future. How often must not the new initiate to the hemodialysis regimen, at least subconsciously, identify with the Shakespearean lament: "Tir'd with all these, for restful death I cry (*Sonnet LXVI*)."

REFERENCES

1. King S: Social psychological factors in illness, in Levine S, Reeder L, Freeman E (Eds): The Handbook of Medical Sociology (2nd ed), New Jersey, Prentice-Hall, Inc., 1972, pp 129–147, p 138
2. Shontz F, Fink S, Hollenback C: Chronic physical illness as a threat. Arch Phys Med Rehab 41:143–148, 1960, pp 143–144
3. Coe R: The Sociology of Medicine. New York, McGaw-Hill, 1970, p 71
4. Coe R: The Sociology of Medicine. New York, McGraw-Hill, 1970, p 73
5. Viederman M: Adaptive and maladaptive regression in hemodialysis. Psychiatry 37:68–79, 1974, p 68–69
6. Rajapaksa T: Maintenance hemodialysis: How to help patients cope. Medical Times, 107:86–92, 1979, pp 86–88
7. Beard BH: Fear of death and fear of life. Arch Gen Psychiatry 21:373–380, 1969, p 373
8. Levy NB: Psychological problems of the patient on hemodialysis and their treatment. Psychother Psychosom 31:260–266, 1979, p 260
9. Friedman EA, Goodwin NJ, Chaudhry L: Psychosocial adjustment to maintenance hemodialysis (part 1). NY State J Med 70:629–637, 1970, p 634
10. Kaplan De-Nour A, Czaczkes JW: Professional team opinion and personal bias—a study of a chronic hemodialysis unit team. J. Chron Dis 24:533–541, 1971, p 534
11. Hurley R: Poverty and Mental Retardation. New York, Random House, 1969, p 133
12. Haggerty RJ: What type of medical care should be offered to the urban poor? in Norman JC (Ed): Medicine in the Ghetto. New York, Appleton-Century-Crofts, 1969, pp 251–259, p 253
13. Kosa J, Zola I, Antonovsky A: Health and poverty reconsidered, in Kosa J, Zola I, Antonovsky A (Eds): Poverty and Health. Cambridge, Mass., Harvard University Press, 1969, pp 319–339, p 322

14. Marston MV: Compliance with medical regimens: A review of the literature. Nurs Res 19:312–322, 1970, p 317

15. Lurie G, Austin R. New etiol. tests for PAS in urine: Report on the use of phemistix and problems on long-term chemotherapy for tuberculosis. Br Med J 1:1679–1684, 1960

16. Bergman AB, Werner RJ: Failure of children to receive penicillin by mouth. N Engl J Med 13:1334–1338, 1963

17. Morrow R, Rabin DL: Reliability in self-medication with isoniazid. Clin Res 13:362–370, 1966

18. Maddock RK: Patient cooperation in taking medications. JAMA 199:169–172, 1967

19. Mohler DN, Wallen DC, Dreyfus, EG: Studies in the home treatment of streptococcal disease: I. Failure of patients to take penicillin by mouth as prescribed. N Engl J Med 252:1116–1118, 1955

20. Davis MS: Variations in patient's compliance with doctor's orders: Analysis of congruence between survey responses and results of empirical investigations. Psychiatry in Medicine 2:31–55, 1971, p 35

21. Meldrum MW, Wolfram JG, Rubini ME: The impact of chronic hemodialysis upon the socio-economics of a veteran patient group. J Chron Dis 21:37–52, 1968, p 49

22. Marston MV: Compliance with medical regimens: A review of the literature. Nurs Res 19,4:312–322, 1970, p 318

23. Bonnar J, Goldberg A, and Smith JA: Do pregnant women take their iron? Lancet 1:457–458, 1969

24. Ireland HD: Outpatient chemotherapy for tuberculosis. Am Rev Respir Dis 82:378–383, 1960

25. O'Brien ME: Hemodialysis and effective social environment: Some social and social psychological correlates of the treatment for chronic renal failure. Unpublished doctoral dissertation, Washington, DC, The Catholic University of America, 1976, pp 122–123

26. Pritchard M: Reaction to illness in long term hemodialysis. J. Psychosom Res 18:55–67, 1974, p 55

27. Mlott S: A psychologist's view of the renal dialysis patient. J AANNT 6:25–31, 1979, p 29

28. Gentry WD, Davis GC: Cross-sectional analysis of psychological adaptation to chronic hemodialysis. J. Chron Dis 25:545–550, 1972, p 548

29. Abram HJ: The psychiatrist, the treatment of chronic renal failure, and the prolongation of life, II Journal of Psychiatry 126:43–52, 1969, pp 45-46

30. Reichsman F, Levy NB: Problems in adaptation to maintenance hemodialysis. Arch Intern Med 30:859–865, 1972, p 861

31. Landsman M: The marginal man, reprinted from Annals of Internal Medicine, Feb 1975, Dialysis and You, Summer, 1977, pp 12–13

32. Duff R, Hollingshead A: Sickness and Society. New York, Harper and Row, 1968

33. King S: Social–psychological factors in illness, in Levine S, Reeder L, Freeman H (Eds): The Handbook of medical Sociology. Englewood Cliffs, N.J., Prentice-Hall, Inc., 1972, pp 129–147, p 141

34. Mechanic D: Medical Sociology. New York, The Free Press, 1968, p 63

35. Kaplan De-Nour A, Czaczkes JW: Personality factors in chronic hemodialysis patients causing noncompliance with medical regimen. Psychosom Med 34:333–344, 1972, p 337
36. Reichsman F, Levy N: Problems in adaptation to maintenance hemodialysis. Arch Intern Med 30:859–865, 1972, p 864
37. Dansak D: Secondary gain in long-term hemodialysis patients. Am J. Psychiatry 129:128–129, 1972
38. Abram H: The psychiatrist, the treatment of chronic renal failure and the prolongation of life II. Am J Psychiatry 126:157–167, 1969, as cited in Dansak D: Secondary gain in long-term hemodialysis patients. Am J. Psychiatry 129:128–129, 1972, p 128
39. Anger D, Agner D: Motivation and adjustment levels of hemodialysis patients. Dialysis and Transplantation 4:49–50, 1975, p 49

─── 7 ───
A Typology of Long-Term
Adaptation to Hemodialysis

*Even as the stone of the fruit must break, that its heart may
stand in the sun, so must you know pain.*

Gibran
The Prophet

Sometimes you don't feel like getting out of bed in the morning, like facing all
this one more time. But you can't think just about yourself. Even though it hurts,
and sometimes it hurts a lot, you have to go on. I guess that's what being a survivor
is all about.

—ESRD patient, 10 years on dialysis

Accepting long-term adaptation to end-stage renal disease and a dialysis regimen involves pain and also involves courage as the above patient comment
so poignantly illustrates. For some, fatigue and physical deficit make even
the ordinary morning rituals associated with rising, something stressful and
problematic. For others, the continued reorganization or modification of
life activities in relation to the therapeutic regimen stimulates the emotional
responses of anger and frustration. For most patients who have dealt with
their condition for longer periods of time, the survivors, an "other-direct-
edness" is found to be a predominant theme involved in life attitudes and
behaviors. Through focusing not upon their own illness condition but rather
upon the needs, cares, and interests of others in their social-interactional
environment, these dialysis patients notably overcome a "sickness" mode
of functioning and operate as productive and contributing members of society.

150

"Long-term" adaptation to hemodialysis is difficult to explore and define at present. As noted earlier, hemodialysis only came into fairly widespread use in major medical centers during the late 1960s and early 1970s. Thus, the majority of survivors or "long-term" patients generally present a history of somewhere between 9 to 12 years "on the machine." (There are, of course, some exceptions. One patient in the present study had a history of 14 years on hemodialysis when he expired.)

Among the patient group in the present research, the mean length of time on dialysis at the time of third interview was 8.5 years. The modal time was 9 years, with a range distribution of 7 to 11 years on the machine. The modal age range of the patient group at T3 (N = 33 patients) was within the 40–49 year category. Eighteen (54.47 percent) of the patients were male and 15 (45.57 percent) were female; and 18 (54.5 percent) of the patients were married.

In this chapter both quantitative (structured interview) and qualitative (focused interview; discussion) data are presented to describe changes over time that took place among the study respondents. These data reflect alterations in both physical and psychosocial dimensions of patient functioning. Among the psychosocial elements evaluated are interactional behavior, quality of interaction, alienation, support of significant others, and compliance behavior. In conclusion, a typology of patient adaptation to long-term hemodialysis is presented.

THE SURVIVORS

In discussing prolonged survival on hemodialysis, Lundin[1] reports that if "survivors" are considered as those patients living longer than 5 years, then certain group attributes may be identified. It is suggested that patients in the early death group (under 5 years) include "the older patient (50 years), usually with a longer history of renal disease and hypertension. Also included are patients with diabetes mellitus, uncooperative, emotionally-disturbed patients, and patients subjected to inadequate dialysis." The survivors or late-death patients (over 5 years) "tend to be younger, with a shorter predialytic illness, with minimal damage to other organs and no progressive systemic disease."[1] In a prospective study of two groups of hemodialysis patients conducted over 24 months, Foster and McKegney discovered that a significantly higher mortality rate found in one group "seemed to be associated with its patients having a much greater degree of psychological disturbance, both in the past and at the time of study entry."[2] Lindner and Curtis make the argument that "the social and psychological problems experienced by dialysis patients are just as important as the medical ones, and probably represent the leading causes of morbidity."[3]

In the present study, data on selected physical and psychosocial variables were compared for the survivors versus those patients who expired during the course of the study. At the time of second and third interviews, patients were asked to report and describe any notable physical problems that they had had during the previous 3-year period. Among the patients who expired between T2 and T3 interviews, some of the conditions included arthritis, bone problems (calcium deficiency), infected fistula, infected graft, pneumonia, neuropathy, and heart attack. Somewhat similar problems, though fewer, were reported at T2 by the present survivors (survivors at T2 and T3). These reports included such conditions as diabetes, neuropathy, arthritis, pulmonary edema, pancreatitis, and heart problems. The list of physical problems presented by the survivors at the time of third interview (T3) was basically the same. Although a number of both survivors and patients who expired listed psychosocial variable changes such as death of a spouse or sibling, there was no notable difference on these factors between the groups.

In regard to psychosocial variables, measured quantitatively, comparisons of 3 groups of patients—those who expired between T1 and T2, those who expired between T2 and T3, and those surviving at T3—revealed significant differences. In regard to social interaction as measured at T1, those patients surviving at T3 had the highest mean scores. This group of survivors reported the greatest amount of social support from family and friends and the most positive quality or satisfaction in their social interactions. The survivors at T3 also reported the greatest degree of secondary gain stemming from the experience with ESRD and dialysis. The patients who expired early on, i.e., between T1 and T2, had demonstrated the greatest degrees of alienation in the T1 interview.

Similar patterns were observed when analyzing the T2 interview data. Patients who expired between T2 and T3 had showed greater degrees of alienation; the T3 survivors reported once again more positive social interaction and quality of interaction, and a greater amount of social support.

Overall, while descriptive patient self-reports of physical and psychosocial problems did not show large differences among the three study patient sub-groups (those who expired preT2, those who expired pre-T3, and survivors at T3) overtime, analysis of quantitatively-measured variables revealed the survivors to be less alienated, more active socially, possessed of more social support systems, and more positive about the quality of their social interactions with family and friends. Comments of several of the survivors illustrate their ways of coping with long-term maintenance dialysis. A middle-aged female patient, on the machine approximately 9 years, observed,

> At first I was really depressed. I'd say, "Why did this happen to me?" I didn't want to come in on this machine. After they [the caregivers] got the lines all hooked up they'd come over with their "speech." They had to wait until I was on [the

machine] because if they did it before, I might *not* go on. Then, one day, I just decided that this wasn't any way to live, so I started doing things—getting active. I think that's being a survivor.

Another long-term female patient commented,

It took me about a year and a half to adjust to dialysis. You've got to live one day at a time. At first I used to just think about coming in to dialysis—and nothing else—all day. But I had to get over that and begin to live. Now I think about all the things I have to do at home and only think about this [dialysis] when it's just time to come.

Finally, a young male patient with an active professional life noted, "You learn to cope; that's all. I live as independent a life as anyone. I live alone and take care of myself. Sometimes I work 70 hours a week. You learn to survive."

Several caregivers also commented on patient survival. One therapist asserted, "The ones that adjust—that don't fight dialysis—last longer than the one that never do. Some resist coming. They say, 'I'd rather die than keep coming here,' and sometimes they do die." Along the same thought a dialysis staff nurse stated, "Survival all has to do with self-care and acceptance. Some never accept dialysis and never will, and then they do real poorly. They may not make it." A dialysis unit social worker expressed it this way:

Patients have to accept dialysis. If they agree to the treatment and if they decide, yes, they want to be on dialysis, then somehow they have to get integrated into a new life. I think it is the only way they are going to survive. If they can somehow come to terms with it in an emotional kind of way, then they can manage the physical and the dietary and all the rest of it. But they have to somehow accept on an emotional level, that this is how it's got to be if they're going to be a survivor.

PHYSICAL CHANGES OVER TIME

A number of physical changes may be observed in most long-term dialysis patients; conditions such as infection, atherosclerosis, ostedystrophy, peripheral neuropathies, and other cardiovascular system problems are common. In a study of physical complications among 1049 hemodialysis patients, Hirschman et al discovered that vascular access problems accounted for 26 percent of the group's hospital admissions; 16 percent of hospitalizations were associated with cardiovascular-related problems: myocardial infarction or congestive heart failure, pericarditis, hypertension; and 11 percent were due to infection.[4] Czaczkes and Kaplan De-Nour, in discussing physical problems in chronic hemodialysis, specifically note such conditions as congestive heart failure; pericardial disease; hypertension; anemia; thrombocytopenia; gastrointestinal complications, e.g., liver disease and hepatitis;

nervous system disorder, e.g., peripheral neuropathy; and metabolic and skeletal system problems.[5]

Neuropsychological evaluations of dialysis patients have revealed abnormalities that led the investigators to suggest that "standard hemodialysis strategies are not able to prevent pathological changes in the brain of uremic patients."[6] A condition labeled "dialysis dementia" has been linked to the deteriorative physiological changes associated with long-term dialysis treatment.[7] Ziesat et al assert that while various factors have been discussed as possible causes of dialysis dementia (e.g., a virus, dopa washout, hypoglycemia, asparagine deficiency, accumulation of drugs, and accumulation of heavy metals), the possible etiology receiving the most widespread support has been aluminum intoxication.[8] These authors suggest that rather than being an "all-or-none phenomenon," it is more likely that dialysis dementia is "distributed throughout the entire population of dialysis patients, such that most dialysis patients would show a mild or moderate degree of dementia, while smaller numbers of patients would show either severe symptoms or no symptoms at all."[8]

As explained previously, all study patients were asked to describe physical problems that had occurred since the preceding interview. A summary profile of these problems includes infected fistula, infected graft, septicemia, "blackouts," "fluid overload," neuropathies, ulcer, chest pains, seizures, hypertension, pneumonia, arthritis, and angina. The spouse of one long-term dialysis patient discussed her husband's occasional irritability, lapses in memory, and confusion, and commented, "The doctors call it dialysis dementia. Sometimes he's difficult, but it's a lot easier to deal with when I know that it comes from the dialysis." Most of the survivors in the present study reported that they had had "ups and downs" physiologically over the course of their dialysis history. Periods of illness associated with physiological problems related to ESRD and dialysis, however, appeared to be viewed by the patient group as an expected part of the renal failure experience.

PSYCHOSOCIAL CHANGES OVER TIME

Long-term adaptation to any serious chronic illness condition may be considered to include certain psychosocial as well as physiological changes. Watkins, in discussing psychosocial functioning, insists that undergoing maintenance hemodialysis is an entirely new way of living never experienced previously, and comments, "Ironically, as the threat of death becomes less and less, the dehumanizing quality of life caused by complete dependency on a machine tends to erode human identity."[9] In discussing the major psychosocial deficits of maintenance hemodialysis patients, Procci asserts that "these deficits appear to be most pronounced in the long-term patient," and adds, "Social functioning, which includes such areas as interpersonal relationships and vocational and avocational pursuits, is most drastically affected by the chronicity of dialytic treatment."[10]

Patients in the present study group were evaluated quantitatively for changes over time in regard to selected psychosocial variables. Findings indicated that between the first two interview periods (T1 and T2) the perceived quality of social interactions declined. This finding is consistent with literature suggesting a deterioration in the quality of life for the long-term hemodialysis patient.[11,12] Alienation increased moderately over the 3-year period. This finding is in agreement with Sorensen's report that after a period of time on dialysis, the novelty of the condition tends to wear off and family and friends pull back, with the result that the patient often becomes socially isolated.[13] Interactional behavior, however, has reported to increase slightly at T2. At the time of the third patient interview (T3), interactional behavior and quality of interaction were both found to decrease moderately and alienation was again increased.[14]

SUPPORT OF SIGNIFICANT OTHERS OVER TIME

Related to psychosocial problems associated with ESRD, and particularly to the unique circumstances of the chronic hemodialysis patients who live dependent upon a machine, continued support of significant others is a most important element in fostering acceptance of the condition and the prescribed regimen.[14] Kossoris asserts that renal patients become very dependent upon their family members and that this involvement and dependency increases as the disease becomes more severe.[15] In discussing adjustment and rehabilitation, Oberley and Oberley cite the words of Dr. Glenn Haswell, a physician who at the time had been on dialysis for over 4 years. Haswell pointed out the importance of the support of significant others for the dialysis patient as time goes on, and asserted, "If the patient's family and associates view him as a chronically ill, dependent person, this can adversely affect his physical as well as emotional and mental adjustment to his life on dialysis."[16]

Dimond, in a study of social support and adaptation to maintenance hemodialysis, found that there was a positive association between family support and patient morale.[17] Kaplan De-Nour, Czaczkes, and Lilos, evaluating chronic hemodialysis caregivers, suggested that "a team's opinion, and especially the agreement or disagreement in their actual expectations, influence patients' behavior."[18] The topic of the support of significant others (conceptualized in terms of primary system members, i.e., family and friends, and secondary system members, i.e., caregivers) was examined in some detail for long-term patients in the present research. In regard to changes over time, patient reports revealed that during the course of the study support of caregivers decreased; family support, however, increased slightly. In looking specifically at the relationship between social support and psychosocial adaptation for the study population, it was found, as anticipated, that moderately strong and significant positive correlation between family support and patient interactional behavior and the quality of interaction existed at the time of initial (T1) and follow-up (T2, T3) measurements. Inverse relationships between social support and alienation were evidenced; the more positively patients perceived the support of significant others, particularly primary system members, i.e., family/friends, the less was alienation evinced.

Primary system correlations remained moderately strong and statistically significant at follow-up (T2); however, a pattern of weakening over time was discerned. This finding is consistent with theories that focus on the strain placed upon interactional relationships between patient and family/ friendship groups and predict its increase over the course of a long-term illness.[19-21] Also relevant to this notion of tension in family relationships, or as King describes it, the "disruption" of interactional patterns, [22] is the decrease over time in strength of the relationship between quality of interaction and the support of family and friends. In discussing the overall psychosocial adjustment of the hemodialysis patient, Schowalter et al report that many studies have questioned the "quality of life" for both patients and their families.[23]

The inverse relationship of alienation scores (those of the total scale as well as those for the subscales of powerlessness, normlessness, and social isolation) with the measure of family/friends support at T1 and T2 demonstrates that when stronger family support is present, alienation is decreased; the association appears to weaken, however, over the 3-year interval.[24] This finding is consistent with the assertion of Hampers and Schupak, who point out that a considerable degree of social isolation occurs if interpersonal relationships with the patient's family and friends are significantly altered in the course of dialysis treatment for renal disease.[25]

LONG-TERM COMPLIANCE WITH THE TREATMENT REGIMEN

Because of the complexity of the treatment regimen associated with hemodialysis, compliance, particularly in regard to dietary and fluid restrictions, is difficult to achieve. Such dietary compliance is even more difficult to maintain over a long period of time; however, as noted by Meldrum et al, dietary as well as overall treatment regimen compliance is an essential factor in chronic hemodialysis patients' survival.[26] In a study on the stress involved in adapting to the dialysis regimen, Joel and Wieder found that all their study patients had problems in adhering to dietary restrictions and most cheated frequently.[27] Gelfman and Wilson assert that dialysis patients often go on fluid and dietary binges, which may endanger their lives.[28] Suicidal gestures, therefore, are often discussed in relation to the chronic dialysis patient. Armstrong commented that up to half the patients in a dialysis center may be noncompliant at a given time and reported that in one study, 8 of 10 patients who died had been identified as noncompliers.[29] Haenel et al suggest that a reason for high rates of suicide among maintenance dialysis patients relates to the fact that "the patient on regular dialysis has ways and possibilities to end his life that other patients do not." They note that "hyperkalemia is a common cause of death in dialysis patients and easy to produce intentionally."[30] As well as being associated with suicide, occasions of gross dietary noncompliance may also be viewed as related to quality of life and/or basic need-gratification for the long-term patient. Procci notes that "severe abuses of the hemodialysis diet may serve an adaptive function by allowing at least some gratification in one area of the lives of these patients who suffer severe depression."[31]

Initial compliance with the dialysis regimen and analyses of significant correlates have been described quantitatively in Chapter 6. It was observed, however, that patients in this long-term study group frequently initiated changes in their compliance behavior. These changes reportedly related to perception of one's own bodily needs or individual ability to cope emotionally with the severely restricted dietary and fluid regimen. In the following section, long-term compliance is discussed utilizing examples of quantitative data reflective of behavior changes, as well as qualitative data from patient comments about their compliance behavior over time.

COMPLIANCE BEHAVIOR CHANGES OVER TIME

Patients' compliance behavior as evaluated quantitatively by a 7-item scale measuring such factors as adherence to dietary and fluid restrictions, taking of prescribed medications, and attendance at regularly scheduled treatment sessions, was found to change significantly over time. Overall, compliance behavior trends appeared to increase both at T2 and T3 for the panel study group (survivors at T3). While there is little in the literature documenting the association between compliance behavior and length of time on a particular therapeutic regimen, the discussions that do exist appear mixed. Blackburn's findings dispute such a linkage.[32] Hartman and Becker reported that dialysis patients who had been on dialysis longer adhered better to all aspects of the prescribed regimen; they suggested that their findings were related to the patients' improved ability to cope with the complex treatment program after a longer period of experiencing the regimen, and they also note a possible association between compliance and survival.[33]

When the individual patient turnover was measured in the present study and cross tabulations calculated for high and low compliers at T1 and T2, much change was found. Of the 34 patients (54 percent) identified as low compliers at T1, 18 remained so at T2, but 16 patients increased in compliance at T2. Of the 29 patients (46 percent) identified as high compliers at T1, 14 were found to fall in the category of low compliers at the T2 interviews. (Categories of high compliance and low compliance were established by dichotomizing the hemodialysis compliance behavior scale on or about the median.)

While in some instances increase in compliance may be related to better ability to cope with a complex treatment regimen over time, reported decrease in compliance behavior might additionally be associated with the nature of the variable itself. The hemodialysis patient most likely learns to "manipulate the system," particularly in regard to dietary, fluid and medication prescriptions. Thus certain of the lower compliance scores found at the time of reinterview may be related to a decrease in "ritual" compliance as conceptualized in terms of strict adherence to the total treatment regimen. The patient may, in fact, have initiated over time a "reasoned" type of compliance behavior more particularly suited to individual needs. Further empirical investigation might provide grounds for re-definition of the concept of compliance, especially for the chronically ill patient.[34]

COMPLIANCE BEHAVIOR AND SOCIODEMOGRAPHIC VARIABLES*

To determine the possible influence of sociodemographic variables on patient compliance, mean scale responses were evaluated for age, sex, marital status, race, education, occupation, type of household, and length of time on dialysis at both T1 and T2. Although differences were noted in the mean responses at T1, none was statistically significant. Haynes, in discussing the determinants of compliance, suggested that little empirical evidence exists to support an association between sociodemographic characteristics and compliance.[35] One theory advanced to explain such lack of evident linkage notes that most patients studied relative to compliance have already entered the health care system and are thus a biased sample; that is, persons who have never sought health care or who have dropped out of a program have been missed in the sample population evaluated.

---

*The text of this section appeared originally in O'Brien ME: Hemodialysis regimen compliance and social environment: A panel analysis. Nurs Res 29: 250–255, 1980, p 253. Reproduced by permission of Nursing Research.

Patients' response differences at T2, however, approached or attained significance on certain scales. Relative to occupation, unskilled workers reported the lowest compliance scores and professionals, the highest. Also notable was the variable, type of household, with subjects who live alone demonstrating the least compliance behavior and those living with adults and children, the most positive. The latter finding is consistent with the report of Meldrum et al, who noted that irrespective of all other factors, dialysis patients with families including several children maintained better than average adaptation.[36] Finally, on marital status, statistically significant differences were found among respondents' reports, with patients who had never been married scoring lowest in terms of compliance with the dialysis regimen. This finding agrees with Hartman and Becker's claim that the less-adherent dialysis patient more often is young and unmarried.[33] No explanation can be offered for the highest mean compliance score reported for divorced subjects.

SURVIVORS AND COMPLIANCE BEHAVIOR

When comparing mean compliance scale scores for the 3 study patient groups in terms of mortality—those patients who expired between Interview 1 (T1) and Interview 2 (T2); those patients who expired between Interview 2(T2) and Interview 3 (T3); and those patients surviving at Interview 3 (T3)—interesting and somewhat unexpected patterns of behavior were discovered. Statistically significant differences appeared between all 3 groups, with those patients who expired earliest (between T1 and T2) demonstrating the *highest* compliance behavior, and those patients surviving at T3 reporting the *lowest* compliance with the therapeutic regimen. This finding might be interpreted as reflecting the earlier discussion of "ritual" versus "reasoned" compliance. The hemodialysis regimen compliance scale used in the study, as noted, measured "ritual" compliance with the dialysis regimen. Most long-term patients reported in their open-ended qualitative interviews that they did not stick strictly to the fluid and dietary restrictions of the regimen. The majority of patients did, however, place importance upon such factors as attendance at scheduled treatment sessions and the taking of prescribed medications. This finding is partially supported by the research of Kirilloff, who reported, "Statements of long-term survivors indicated that both strict adherence to the prescribed regimen and careful extension of limits beyond those 'officially' prescribed may be a characteristic of at least some survivors of long-term dialysis."[37]

Virtually all long-term study patients were open in admitting to and discussing their various "dietary indiscretions." From the qualitative data gathered in these discussions, several themes emerged related to why patients do not comply with their prescribed regimen. Three categories of reasons reported were identified as concerning (1) physiological/ nutritional needs, (2) psychological/normalcy needs, and (3) social/interactional needs.

Physiological/Nutritional Needs

A number of study patients reported that their prescribed dietary regimen simply did not provide them with enough food "to survive" or to supply the energy to carry out daily activities. One 42-year-old male patient, on dialysis 11 years, commented, "At first I followed that diet religiously but I just found myself getting weaker and weaker. I found that by eating more I felt better. I don't go way off [the diet] though—only within the bounds of what I know I can do." Another male patient, who had been on the machine 9 years and was working full-time, noted, "I know what I can and can't do. I cut back on some foods and fluids but I never stick to the diet. I couldn't eat that diet and have the strength to work too. I knew that right from the beginning." This patient added, however,

> You have to know the things you can't eat, like potassium and salt.

A young man who maintained a successful professional career summed up his feelings this way,

> "You can't live strictly on that diet. It's not an intelligent diet. The diet restricts you more than anything else. They told me at first that I would never eat in a restaurant again. But you can't just eat anything you want, either. I know how to "cheat" according to my chemistries. I'm like "weight-watchers"—I practice "intelligent cheating."

Psychological/Normalcy Needs

Certain patient responses in regard to reasons for cheating on the dietary regimen were linked to personal identity/normalcy needs. As one female patient complained, "That diet isn't normal. You feel you can't ever eat anything you want. Sometimes you feel like you've just got to have some 'junk food'—the stuff other people eat. You get so tired of it [the diet], sometimes I just feel depressed and get myself something with salt." A male patient reinforced the above statement, commenting, "There are so many things that you can't have, that you don't feel like a normal human being. Sometimes I just eat things like potatoes. I know I shouldn't eat too many, you know, but I just love potatoes. I don't cheat much on nuts and stuff like that." Another male patient commented, "The diet does not give you what you need to have to live a normal life. Ninety percent of the patients who have died in this place over the past five years have died because of that diet— because of malnutrition. You can't live on that diet. It will kill you." Finally, one long-term female patient described a specific instance of noncompliance this way:

Last week I went shopping and I had to walk about three to four blocks in that really bad wind. I got so cold I thought I would die. So, I stopped at MacDonald's, I was so cold, and ordered hot chocolate and a cheeseburger. I never do anything like that but I just had to and then I said, "Oh God, please don't let me die!" But I just had to have something.

Social/Interactional Needs

Several patients interviewed discussed the importance of sharing a meal or a particular kind of food with others as a social event. One young female commented, "Once in a while you've got to go out for beer and pizza with your friends. You can do it if you watch [what you eat] the day before. And then, too, you only have one piece of pizza and one glass of beer." Another female patient noted, "On a holiday like Christmas or Easter, when you're with your family, you've got to have just a 'little taste' of things—just don't be too carried away." A male patient admitted, "On the weekends, when I run with my friends, that's when I do go over. Sometimes I'll binge—really eat and drink a lot of fluids." Several patients mentioned violating their dietary prescriptions when "out to dinner" with friends but, as one female patient remarked, "I'll be honest, I cheat but I know when to stop. If I go out and have a nice dinner, I watch it the next day. And in the summer, I don't eat as heavy so the fluids aren't too bad."

Although, as mentioned earlier, most of the study patients did admit to dietary indiscretions, the majority of these respondents also reported that they had learned their own "limits" and would never grossly fail to comply to such a degree as to endanger their lives. One patient commented that he had learned his lesson the "hard way"; in his own words, "I 'binged-out' on fluids once and got pulmonary edema. That really scared me. I thought I was going to die—thought I was gone. I never want to go through that again." A hemodialysis therapist interviewed in the study commented, "I mean nobody totally sticks to that diet, none of the patients do, but they know how much they can eat and how fast they can dialyze it off."

A registered nurse who had been working with dialysis patients for approximately 6 years expressed her understanding of patient noncompliance this way:

Well, the way I feel is, you've seen those renal diets, and they are really horrible. I mean go and tell somebody, "Well, this is what you have to eat for the rest of your life." Maybe there's some people, they don't care, but not most. You know, in our society, eating is a really important thing, a socialization type thing. So, it's true, people need to stick to their diet, to their food restriction to some extent. But I think we need to work with them as individuals. Like, maybe someone drinks a little more and it doesn't seem to hurt them that much. Of course, you know, if he comes in [to dialysis] and he's gained 20 pounds so he can't even breathe, then that's another issue. We need to work with the individual.

It should be noted that contemporary renal diets tend to be more palatable due to the increase of individualized patient dietary planning and counseling.

Perhaps one of the most significant aspects of patient compliance/noncompliance with the dialysis dietary regimen has to do with the symbolic meaning of food. Food is a source of sustenance, pleasure, and physical satisfaction of hunger. The preparing and serving of food often involves a loving, caring relationship, providing spiritual and psychological nourishment for the one being served and the one serving. The sharing of a meal with another adds a social-interactional dimension to the concept of food.

Certain foods such as desserts and beverages like coffee, tea, or alcohol provide special social-symbolic satisfaction; in some cases they may be related to rewards and celebrations for fulfilling life accomplishments. For the chronically ill dialysis patient, the restriction of food, especially of certain foods associated with greater sensory and psychological pleasure, adds yet another stressor to an already heavily laden therapeutic regimen. Dietary noncompliance is a complex and frustrating phenomenon, not totally understood by either patient or caregiver. Frequently, patient "binges" or instances of gross noncompliance may be interpreted as a desperate attempt to be "normal," efforts to deny the illness condition in the magical hope that it will go away. Sometimes such noncompliant behaviors can be seen as expressive of an overwhelming sense of frustration and fatalism—a sort of "last good meal." Whatever the psychological or sociological catalyst of the behavior, serious and continual noncompliance with dietary regimen has grave physiological ramifications for long-term dialysis patients and needs continual assessment and evaluation.

A TYPOLOGY OF ADAPTATION TO LONG-TERM HEMODIALYSIS

A primary purpose of this longitudinal study with maintenance dialysis patients was an attempt to identify a composite or several composites of long-term life adaptation to ESRD and the hemodialysis treatment regimen. At T3, six years following initial contact, patients were visited several times in order to conduct open-ended qualitative interviews dealing with long-term dialysis adaptation. Significant others (family members and friends) were interviewed also.

As focused interviewing and observation were being carried out, an attempt was made to cluster individual patient characteristics, attitudes, and

Table 7-1.
Typology of Patient Adaptation to Long-Term
Hemodialysis: Themes (Attitudes and Behaviors)

Type I: "Career" Dialysis Patient	Type II: "Part-Time" Dialysis Patient	Type III: "Free-Lance" Dialysis Patient
Passivity	Autonomy/control	Vascillates between passivity and control
Dependency (regarding care)	Self-care (to the degree possible)	Vascillates between self-responsibility and dependence
Attitude toward dialysis: life centers on hemodialysis	Dialysis viewed as a part-time job: life centers on work, family, friends	Ambivalent re dialysis regimen and necessitated modification of lifestyle
Likes coming to the dialysis unit	Tolerates coming to the dialysis unit	Dislikes coming to the dialysis unit
Much interaction with other dialysis patients and caregivers	Minimal interaction with other dialysis patients and caregivers	Vascillates in amount of interaction with other dialysis patients and caregivers.
Engages in very little, if any, meaningful nondialysis work or life activity	Engages in much meaningful nondialysis work and life activity	Involvement in meaningful nondialysis work and life activity varies with mood
Practices "ritual" or modified compliance	Practices "reasoned" noncompliance	Practices modified compliance or gross noncompliance
Adopts the "sick" or "disabled" patient role	Adopts the "well" patient role	Adopts the "chronically ill" patient role
Lives in the "sick" dialysis world	Lives in the "well" dialysis ("normal") world	Marginal to dialysis world and normal world

behaviors into meaningful categories that might have some theoretical import in evaluating the life career of the chronic dialysis patient. A number of the emerging categories related to such topics as sick-role behavior, compliance with the treatment regimen, attitudes toward dialysis, interaction with dialysis patients and caregivers, social relationship, family role behaviors, and work activities. As these categories became more clearly defined (through comparing the properties of the individual patient and behaviors comprising a category), the analysis moved on to the stage of delimiting the construct or theory (Glaser and Strauss[38]). Through this process, a typology of adaptation to long-term hemodialysis emerged consisting of three distinct types of dialysis patient attitudes and behaviors. The individual patient types are: Type I: The "Career" Dialysis Patient; Type II: The "Part-Time" Dialysis Patient; Type III: The "Free-Lance" Dialysis Patient (see Table 7.1).

Type I: The "Career" Dialysis Patient

Long-term hemodialysis patients categorized as Type I or "career" dialysis patients were those whose lives appeared to be almost totally centered on hemodialysis and the therapeutic regimen. These patients tended to be more dependent on caregivers for their treatment and frequently expressed the opinion that they did not mind and even liked coming to the unit for dialysis. One female patient was observed to come to the outpatient dialysis unit always very well dressed and with her hair attractively arranged. When asked about this, she commented, "This is the only place I come all week. I don't go anywhere else. I come and see my friends; so, I like to get dressed up and look nice. I have a lot of friends out here [at the dialysis unit]." Another older female patient whom caregivers identified as fitting the "career" dialysis patient type was described thus by a dialysis unit social worker:

She is a fairly physically healthy lady whose husband died about 6 or 8 years ago and her life was just meaningless, living in a big house wondering what to do with herself. And boom! Kidney failure. She didn't want to live necessarily, but she didn't verbalize that she didn't want to live. And obviously she got into hemodialysis and it is the highlight of her life. Her entire life revolves around it. She has said, "Dialysis is something for me to live for. My relationships with people here are important. It gives me something to do. It organizes my life. I get out."

The social worker added,

She gets her hair done so that she looks good when she comes in to the unit. She is very concerned about other patients. They have become her friends now. She has developed really very meaningful relationships with both unit staff and the physicians. And it is her reason for living.

Several other caregivers described patients who they felt could be identified as "career" patients. Of a long-term female patient, this comment was made: "She has no family and she really enjoys coming here [to the dialysis unit]. This is the only place she ever comes. She mingles a lot with the other patients and the staff." Another caregiver offered these remarks about a young female patient: "If I were her spouse or child I think I would find it very difficult. Her whole life revolves around these treatments. Dialysis is a 'profession' for her." A dialysis unit head nurse described a middle-aged male patient this way:

He stays in his bedroom when he is between treatments. His only social activity, interaction, is right here in this unit. He comes in about seven in the morning and hangs around until one to one-thirty, because he's probably happy to be around the people here. He knows he can have a cup of coffee after he is dialyzed, smoke a cigarette, and talk to us. That's probably his only interaction.

Type II: The "Part-Time" Dialysis Patient

A number of long-term hemodialysis patients might appropriately be categorized and did, in fact, characterize themselves, as "part-time" patients. Their lives were filled with many interests, concerns, and behaviors not related to ESRD and the hemodialysis regimen. Several patients in this group described the attendance at hemodialysis treatment sessions as like having a "part-time job." One young male patient with a full-time professional career commented,

> Dialysis is like having a second job, a part-time job. I resent it right now because of facing career decisions. Others can work more hours now than I can. I have to manage my schedule for travel; mostly it's just irritating. Also not having the evenings during the week limits your social activities, like going to the theater or taking a course two nights a week. It interferes with social activities.

Another male patient who was working full-time at a semi-skilled occupation remarked, "Dialysis is definitely not the center of my life. It's like a part-time job—the hours I have to come here [dialysis unit]. My friends on the job really help a lot. They treat me like an ordinary person." A female patient, a housewife, the mother of several young children, stated, "Being on dialysis is like having a job. You have to do it 3 times a week. Just like my husband goes to work—it's like another job." Another female patient, also a wife and mother, observed,

> When I leave here [dialysis unit] I like to just forget about dialysis, to forget I'm a dialysis patient. You can spend all your time thinking about dialysis and talking about being a dialysis patient and you can't *do* anything. People don't really want to hear all that—it's better not to talk about it. When I go home, I have my family to think about and I have the house and my church work. I don't think about this place until it's time to come in again.

Type III: The "Free-Lance" Dialysis Patient

Perhaps the most difficult group of patients to categorize and describe are those patients whose behaviors and attitudes regarding ESRD and the dialysis regimen are vascillating. These patients are at one time compliant, at another noncompliant, in terms of the dietary and fluid restrictions, and even, on occasion, in regard to attendance at scheduled treatment sessions. Patients in this category were not overly involved either in dialysis or in other aspects of their daily lives, but appeared marginal to the world of the sick and to the world of the well. When asked to characterize a patient representative of this type of adaption, a dialysis caregiver replied, "He's a 'sometimie' patient. Sometimes he does right and has no problems. But sometimes he doesn't even show up [for a treatment] or comes in way over [fluid overloaded]. He'll go 4 or 5 kilos over and expect us to take it off. He abuses the diet a lot." A male patient interviewed in the study appeared to represent the "free-lance" group by his own comments:

> Sometimes I miss a Wednesday [treatment session], but I do okay. I stick to the diet if I feel like it, but sometimes I just eat what I want—like salt and pepper and all kinds of seasoning. I drink whatever I want but I get "short-winded" if I go overboard. I don't do good with the medicines—some of them make me dizzy and sick at the stomach. I can do for myself if I want—but I don't do much. My wife, she really takes care of me.

One long-term dialysis caregiver described patients of the "free-lance" type this way:

> I think some of these patients really need to just "do their own thing" because they need the control. It gives them control, if they can skip a treatment once in a while or binge on fluids. I have a man who comes into the unit, who finds it fun, like a game. He's proud of the fact that he can skip a treatment sometimes. He'll come in here and he may go on that machine or may not. He may stay 15 minutes, he may stay for his full treatment. But he may, then again, come in here, wander around, smoke a cigarette, and never once get on the machine; and it really frustrates me. I've argued with him. And I've really gotten angry and I felt like shaking him. We have had some good blowouts and afterwards he'll come back in [to the unit] and try to pretend that he was confused—that he didn't know that he did that. But sometimes he can be really compliant too.

The three-dimensional dialysis patient typology describes patterned activities involved in patient "work," whether it be of a career, part-time, or free-lance nature. The business or "work" of being a patient may be carried out successfully in different manners for patients of different characters and personality types. For each group or type, what appeared critical to satisfactory adaptation to ESRD and its therapeutic regimen (and thus to survival) was the acceptance of the illness condition, at least to a degree. Once this was accomplished, the work of lifestyle adaptation and coping could be initiated.

Although the above 3 patient types did emerge from observation and interviewing with long-term maintenance hemodialysis patients, it should be understood that patients may change type (or change certain typical attitudes and behaviors) over the course of a number of years on dialysis. Some patients may obviously adopt some of the attitudes and behaviors related to the "career" type of adaptation early on in their coping with ESRD and dialysis or may revert to these attitudes and behaviors at a later period if psychological problems interfere with activities of daily living. It is conceivable that a patient may change from being a "career" type dialysis patient to a "part-time" type patient as he or she learns better how to cope with renal failure and the lifestyle modification necessitated by dialysis treatment. A patient may also move from the "career" or the "part-time" adaptive type to the "free-lance" mode should the stress of long-term coping with the illness condition prove too discouraging.

In evaluating the T3 panel group of study patients, however, definite consistencies in attitude and behavior were observable as related to type identity. Among the 33 patients involved in the study at T3, 10 patients were identified with the "career" adaptive type, 14 patients were representative of the "part-time" dialysis patient type, and 9 patients fit the "free-lance" mode of adaptation. In order to validate this qualitatively derived typology, quantitatively measured social functioning variables were compared across the 3 identified patient-type groups. Analysis of T1, T2, and T3 data revealed statistically significant differences between groups on several of the major social functioning variables, including family support, interactional behavior, quality of interaction, and alienation. At T3, post hoc analysis determined that the "part-time" group were distinguished significantly from the other two groups in demonstrating a greater degree of interactional behavior and more positive quality of interaction. The "career" group were significantly more alienated not only in terms of overall alienation but also on each of the instrument's three subscales, which individually measure powerlessness, normlessness, and social isolation. Perhaps the most formidable aspect of the above analysis is that the pattern of significant difference not only persists but strengthens over time (T1◊T3), from which one might infer that the patient typology of adaptation firms up and becomes more distinct the longer one experiences the hemodialysis patient role.

Long-term hemodialysis patients, the winners in a continuing battle to provide some semblance of normalcy in their daily lives, must be applauded for their perserverance and courage in the face of physical and psychosocial deficits. Interactional behavior may sometimes be modified in relation to the demands of a time and energy-consuming therapeutic regimen. The quality of such interaction may also change due to a patient's fatigue or inability to maintain a former level of sharing in a relationship. As a result, loneliness and feelings of serious alienation occasionally surface. Compliance with the treatment regimen, particularly in its dietary and fluid limitation aspects, is difficult initially and can become almost intolerable as years of such restrictions are experienced. Overall, long-term hemodialysis patients may truly be said to be survivors, triumphing over the devastating impositions of their illness.

REFERENCES

1. Lundin AP: Prolonged survival on hemodialysis. Int J Artif Organs 4:7–8, 1981, p 7
2. Foster FG, McKegney FP: Small group dynamics and survival on chronic hemodialysis. Int J Psychiatry Med 3:105–115, 1978, p 112
3. Lindner A, Curtis K: Morbidity and mortality associated with long-term hemodialysis. Hosp Pract 9:143–150, 1974, p 149
4. Hirschman GH, Wolfson M, Mosimann JE, Clark CB, Dante ML, Wineman RJ: Complications of dialysis. Clin Nephrol 15:66–74, 1981, p 66
5. Czaczkes JW, Kaplan De-Nour A: Chronic Hemodialysis as a Way of Life. New York, Brunner/Masel, 1978, pp 45–72
6. Gilli P, Bastiani RG, Fiocchi O, Squerzanti R, Tataranni G, Farinelli A: Impairment of the mental status of patients on regular dialysis treatment. Proc Eur Dial Transplant Assoc 17:306–311, 1980, p 306
7. Ware CD: Dialysis dementia. Nephrology Nurse 3:13–16, 1979
8. Ziesat HA, Logue PE, McCarty SM: Psychological measurement of memory deficits in dialysis patients. Percept Mot Skills 50:311–318, 1980, p 313
9. Watkins YB: Rehabilitation risks of ESRD. J Rehabil 1:30–33, 1979, p 30
10. Procci WR: Psychosocial disability during maintenance hemodialysis. Gen Hosp Psychiatry 3:24–31, 1981, p 30
11. Levy NB: Living or Dying: Adaptation to Hemodialysis. Springfield, Ill., Charles C Thomas, 1974
12. McKegney FP, Lange P: The decision to no longer live on chronic hemodialysis. Am J Psychiatry 128:47–54, 1971
13. Sorensen E: Group therapy in a community hospital dialysis unit. JAMA 221:897–901, 1972
14. O'Brien ME: Effective social environment and hemodialysis adaptation: A panel analysis. J Health Soc Behav 21:360–370, 1980, p 365
15. Kossoris P: Family therapy as an adjunct to hemodialysis and transplantation. Am J Nurs 70:1730–1733, 1970
16. Oberley ET, Oberley TD: Understanding your new life with dialysis. Springfield, Ill., Charles C Thomas, 1976, p 57
17. Dimond M: Social support and adaptation to chronic illness: The case of maintenance hemodialysis. Res Nurs Health 2:101–108, 1979, p 101
18. Kaplan De-Nour A. Czaczkes JW, Lilos P: A study of chronic hemodialysis teams—differences in opinions and expectations. J Chron Dis 25:441–448, 1972, p 446
19. Coe R: Sociology of Medicine. New York, McGraw-Hill, 1978
20. Holcomb JL: Social functioning of artificial kidney patients. Soc Sci Med 7:109–119, 1973
21. Parson T, Fox RC: Illness, therapy and the modern urban family. Journal of Social Issues 8:31–44, 1952
22. King SH: Social psychological factors in illness, in Freeman HE, Levine S, Reeder LG (Eds): Handbook of Medical Sociology. Englewood Cliffs, N.J., Prentice Hall, 1972, pp129–147
23. Schowalter J, Ferholt J, Mann N: The adolescent patient's decision to die. Pediatrics 55:97–103, 1973

24. O'Brien ME: Effective social environment and hemodialysis adaptation: A panel analysis. J Health Soc Behav 21:360–370, 1980, p 364
25. Hampers C, Schupak E: Long-term Hemodialysis. New York, Grune and Stratton, 1967
26. Meldrum MW, Wolfram JG, Rubini ME: The impact of chronic hemodialysis upon the socio-economics of a veteran patient group. J Chron Dis 21:37–52, 1968
27. Joel S, Wieder SM: Factors involved in adaptation to stress of hemodialysis. Smith College Studies in Social Work 43:193–205, 1973
28. Gelfman M, Wilson EJ: Emotional reaction in a renal unit. Compr Psychiatry 13:283–290, 1972
29. Armstrong SH: Psychological maladjustment in renal dialysis patients. Psychosomatics 19:169–171, 1978
30. Haenel T, Brunner F, Battegay R: Renal dialysis and suicide: Occurrence in Switzerland and in Europe. Compr Psychiatry 21: 140–145, 1980, p 143
31. Procci WR: Psychological factors associated with abuse of the hemodialysis diet. Gen Hosp Psychiatry 3:111–118, 1981, p 111
32. Blackburn JL: Dietary compliance of chronic hemodialysis patients. J Am Diet Assoc 70:31–37, 1977
33. Hartman E, Becker H: Noncompliance with prescribed regimens among chronic hemodialysis patients. Dialysis and Transplantation 7:978–983, 1978
34. O'Brien ME: Hemodialysis regimen compliance and social environment: A panel analysis. Nurs Res 29:250–255. 1980, p 254 [Permission to quote from Nursing Research]
35. Haynes RB: A critical review of the determinants of patient compliance with therapeutic regimens, in Sackett DL, Haynes RB (Eds): Compliance with Therapeutic Regimens. Baltimore, Johns Hopkins University Press, 1976, pp 26–39
36. Meldrum MW, Wolfram JG, Rubini ME: The impact of chronic hemodialysis upon the socio-economics of a veteran patient group. J. Chron Dis 21:37–52, 1968
37. Kirilloff LH: Factors influencing the compliance of hemodialysis patients with their therapeutic regimen. J AANNT 8:15–20, 1981, p 20
38. Glaser A, Strauss A: The Discovery of Grounded Theory. Chicago, Aldine, 1967

—8—

Summary and Projections for the Future: The Courage to Survive

It was by faith that Abraham obeyed the call to set out for a country that was the inheritance given to him, and he set out without knowing where he was going.

Hebrews 11:8

HEMODIALYSIS: THE LONG-TERM VIEW

In the preceding chapters an attempt has been made to allow the reader to experience vicariously the world of the long-term hemodialysis patient, and those of dialysis patient family members and caregivers as well. One finds overall that the long-term patient, the survivor, whether primarily involved with career, with family, or with the regimen itself, has learned successfully to manage the dialysis patient role.

In evaluating the dialysis patients in their long-term adaptive and survival mechanism, one hears in the survivors' own words of the difficulties associated with ESRD and hemodialysis: the alienation, dependency, the necessitated modification of life styles, and even in some instances of the perceived stigma associated with the illness condition. Patients report the feelings of loneliness that may surface particularly when evaluating their own human condition against that of a healthy colleague or friend. Sometimes the rules of the game may appear terribly unfair and norms of behavior become obscure and confusing, clearly evidencing a case of Durkheimian alienation.

Individual dialysis patients also recall the pain and personal threat caused by the deaths of close friends who were on dialysis, and report the emotional need to "pull back" after such an occurrence. Patients acknowledge the positive benefits of sharing with another who is like themselves, someone who understands and with whom they can compare stressful illness experiences. However, the relationship itself may turn into a stressful occurrence if the dialysis patient friend begins to deteriorate either physically or emotionally. Strong identification with the friend's increasing impairment may result in serious personal threat for the patient companion. Frequently such an experience results in a patient determining to cease participating in ongoing relationships with other dialysis patients.

Yet finally, even while discussing the many problems of long-term dialysis, even while sometimes questioning the quality of their lives and the uncertainty of their futures, a considerable number of patients describe positive aspects of the hemodialysis regimen and express deep gratitude for continued survival.

The "families" of dialysis patients, those who have been identified as significant others, discussed the impact of ESRD and hemodialysis on their lives and on their relationships. Noted particularly are instances of role reversal for a spouse, the modification or reorganization of family social activities, and the confusion and also the support of the children. Although family members frequently alluded to a kind of "burnout"—a periodic need to get away from the problems associated with long-term chronic illness—most supported in a very positive manner the coping of the dialysis patient friend or family member. Overall, family members' attitudes and behavior may be understood in terms of Duff and Hollingshead's conceptualization of centripetal versus centrifugal activity within families facing an illness condition.[1] Dialysis patients' family members reflecting the "centripetal" mode tended to mobilize their resources and draw together to meet the imminent threat to the integrity of the family structure. A "centrifugal" orientation is demonstrated when individual family members withdrew into individual isolation, neglecting the needs of others in the family unit as well as those of the hemodialysis patient.

Hemodialysis caregivers, self-described as needing a great deal of patience and the ability to deal with chronicity and death, spoke of their frustrations and their "burnout" after longer periods of work in dialysis where patients do not "get better" and go home (with the exception of successful kidney transplantation). Coping with gross noncompliance and patient death were noted as their greatest stressors. Also referenced as extremely painful was the occasional patient decision to terminate the treatment regimen. Out of the empirical data generated through open-ended interviews with professional and paraprofessional dialysis unit personnel, a typology of caregiving emerged. The 3 caregiver types identified were the "machine-tender," the "counselor," and the "confidante."

Caregivers identified with the "machine-tender" type generally reported themselves to be primarily concerned with the mechanical and physiological functions of the dialysis treatment procedure. They denied any real involvement with the patient as person. "Counselor" type caregivers verbalized a more holistic approach to their role behaviors, expressing interest in the patients' overall life adaptation to their illness condition. They did not, however, desire any contact with patients outside of the dialysis treatment setting, i.e., they did not want to "take their jobs home with them." Caregivers in the "confidante" type group reported more in-depth patient involvement, asserting that intense relationships often were developed with long-term patients and that social contact sometimes occurred away from the dialysis unit. Of course, an individual caregiver's type might change or be modified to a degree as a result of such variables as increased experience in the work setting or the burnout phenomenon.

In examining the dialysis unit as a social system, one finds a unique sociomedical environment where many ordinary norms of patient–caregiver relationships are modified or openly violated. Patients may vie with professional personnel for even minimal instances of control, as exemplified in their "going-on" and "coming off" rituals. This is often done to shore up one's weakened autonomy or provide some small measure of power over a future destiny. Much patient ambivalence is directed toward the "machine"—that terrifying yet respected technological creation that may both preserve a life and weaken it, that may support or hinder daily activities, and that may open the door to a fulfilling future, or close that same door on a formerly active career. In the hemodialysis unit, where life is supported and life is threatened, numerous ethical dilemmas revolving around continued survival confront the ESRD patient: death brought on through failure to participate in the dialysis procedure; death also a real possibility through gross and continued noncompliance with the therapeutic regimen. Caregivers and patients struggle together with such issues, trying to find appropriate and humane solutions acceptable to all involved.

Although a primary goal of the hemodialysis unit is admittedly technological, much care-giving activity is also directed toward assisting the patients to achieve a holistic life adaptation to their condition. Thus interaction between caregiver and patient may become important. Often long-term and intense relationships develop among and between both patient and caregiver groups, and a complex social life emerges within the physical and territorial boundaries of the dialysis unit system. Caregiver and patient relationships appear frequently to contain affective as well as instrumental elements; the intermeshing at times proves problematic. When the caregiver must assume a posture of authority, positive affect must be suppressed in order that appropriate instrumental behavior be operationalized. Negative sanctioning of patients by caregivers is generally viewed as an expression of instrumental behavior and is frequently related to activities reflecting patient noncompliance with the therapeutic regimen. Negative sanctioning of caregivers by patients tends to fall within the affective realm, with emphasis placed upon care or lack of care between individuals (withdrawal of love).

In evaluating early adaptation to the dialysis regimen, the most notable stressors identified related to such factors as devalued social status, financial concerns, dependency on the "machine" and on caregivers, physiological and psychological fatigue, family role disruption, and uncertainty of the future. Early regimen compliance was found to be fostered through the support of significant others, both "family" members and dialysis caregivers, with the caregivers' influence being most important for patients in the lower socioeconomic status categories. From the patient interview data a "sickness–wellness self-perception continuum" was generated, representative of 3 individual modes of patient self-perception and functioning regarding end-stage renal failure.

The modes of patients' self-perception identified were the "self-perceived sickness mode," the "self-perceived chronic illness mode," and the "self-perceived wellness mode." In the "sickness" mode patients perceive and report themselves to be sick, fighting to alleviate their illness condition. The "chronic illness" mode places primary focus upon learning about and adhering to the regimen, and reorganizing one's life-style to cope with the illness condition. In the "wellness" mode the greater part of the patients' interest and energy are directed toward other life activities and goals; ESRD and the dialysis regimen are secondary. These 3 modes are envisioned as comprising a continuum on which patients' attitudes may be located. Obviously, patients may move back and forth along the continuum in concert with physical or psychosocial stresses that occur in their lives.

The evaluation of long-term adaptation to maintenance dialysis revealed numerous patterns of patient behavior and behavior change over time. In distinguishing the survivors (those patients surviving at T3 interviews) from those patients who expired between T1 and T2 interviews and those patients who expired between T2 and T3 interviews, it was found that overall the survivors (at T3) were less alienated, possessed of some effective social support systems, more active socially, and more positive about the quality of their social interactions with family and friends. A number of physical changes were identified by study patients, such as infected grafts or fistulae, calcium problems, neuropathies, heart problems, arthritis, fluid overload, and fatigue. Psychosocial changes reported were associated with certain experiences, including death of a spouse or sibling, alienation from friends and family, inability to work or carry out family role responsibilities, and decreased quality of interactions with family and friends. Long-term compliance with the treatment regimen was found to be modified by most of the survivors and a distinction emerged relative to a "reasoned" versus a "ritual" type of compliance behavior. Patients' rationales for noncompliant behavior over time were categorized as relating to 3 dimensions of human needs: (1) physiological/nutritional needs, (2) psychological/normalcy needs, and (3) social/interactional needs. Finally, out of the qualitative patient interviews, which elicited data reflective of the life career of the maintenance hemodialysis patient, a typology of adaptation to long-term hemodialysis emerged, consisting of 3 distinct types of dialysis patient attitudes and behaviors. Type I. The "Career" Dialysis Patient; Type II. The "Part-Time" Dialysis Patient; Type III. The "Free-Lance" Dialysis Patient.

"Career" type dialysis patients' lives appear to be almost totally centered upon hemodialysis and the therapeutic regimen. Coming to the dialysis unit for treatment provides an outlet for their social interactional needs, and the development of close relationships with other patients and with caregivers is the norm. In distinction those patients in the "part-time" group lead lives directed primarily toward interests and concerns not immediately related to ESRD and dialysis. Some patients of this type actually describe the required attendance at scheduled hemodialysis treatment sessions as "like having a part-time job." Patients identified with the "free-lance" group are those whose attitudes and behavior regarding ESRD and dialysis are vascillating. They appear marginal both to the world of the sick and the world of the well. Although the above patient adaptive types emerged as reflective of consistent attitudinal and behavior patterns among a group of long-term hemodialysis patients, it is speculated that patients may change type, or change certain of the associated attitudes and behaviors, during the course of a lengthy dialysis career.

Although the material presented in the preceding chapters has focused upon hemodialysis or more specifically long-term adaptation to ESRD and the hemodialysis regimen, it is felt that many of the stresses, problems, and adaptive modes experienced by the maintenance in-center hemodialysis patient may also be relevant for the ESRD patient opting for a different treatment modality. There are at present a number of alternative procedures that may be options for a chronic renal failure patient. These include in-center hemodialysis, home hemodialysis, IPD (intermittent peritoneal dialysis), CAPD (chronic ambulatory peritoneal dialysis), CCPD (continuous cyclic peritoneal dialysis), and kidney transplantation.

ALTERNATIVES FOR THE ESRD PATIENT

In-Center Hemodialysis

At the present writing in-center hemodialysis remains the most widely used treatment modality for end-stage renal failure. Writing in *The New England Journal of Medicine,* February 1981, Gutman et al estimated the number of Americans undergoing maintenance hemodialysis at between 50,000 and 60,000.[2] That number is increasing daily. Relman and Rennie noted that only approximately 13 percent of these patients were receiving treatment at home.[3] New hemodialysis outpatient units, either free-standing or located within a hospital or medical clinic, are continuing to open in order to accommodate the increasing dialysis patient population. Although home hemodialysis is supported and strongly encouraged in a number of areas of the country, many patients refuse this option as they fear the disrupting effects that such home treatment may have on the family. Another reason for refusal is related to the stress of the care-giving responsibility placed upon the dialysis patient's partner, frequently a spouse or close relative.

Presently most in-center dialysis units schedule treatments on alternate days, 3 days a week, with approximately 3–4 hours planned per patient session. Equipment varies, sometimes depending upon the length of time the center has been in operation. The administrator of a recently opened dialysis unit proudly displayed the latest model of artificial kidney, which she described as the "Cadillac" of dialysis machine technology.

Lowrie et al testify that physicians have long been confronted with the problem of how to prescribe the appropriate amount or "dose" of hemodialysis for each patient. They add that, "in clinical practice most patients in a dialysis program undergo treatment for similar lengths of time and with dialysis machines that have similar performance characteristics."[4] Czaczkes and Kaplan De-Nour report that in the past, dialysis centers attempted to find the most workable schedule in order to accommodate larger numbers of patients and promote maximum utilization of equipment and personnel. They note, "It became accepted practice to dialyze a patient either for five to six hours three times a week on a coil type dialyzer or 12 hours twice a week on a parallel flow dialyzer, like the standard Kiil dialyzer."[5] As explained earlier, the time for treatment sessions is presently decreasing, related in some measure to the more sophisticated technology of the procedure itself. It has been noted, however, that

> dialysis has not undergone any true technical or theoretical breakthroughs. The startling improvement in survival must thus be the result of many small technical advances. Among these are probably the switch from shunts to fistula, the increasing use of three rather than two times a week dialysis, the much earlier start of dialysis, and better blood pressure control.[6]

A number of patients in the present longitudinal study described their rationale for choosing in-center rather than home hemodialysis. A female patient, a wife and mother of two teenagers, stated, "When I leave the dialysis unit I just want to forget about this place. I don't want dialysis in my home or around my family. It's enough to have to come in here three times a week without bringing that machine into the house." A male patient commented, "I live alone and it would be too much to do this at home. I couldn't ask my daughter to come over every other day and run the machine." The spouse of a male dialysis patient reported, "They wanted him [the patient] to do this at home but I said no. They tried to teach me about the machine but I was afraid. I didn't really want that responsibility. If something happened to him, I think I'd panic."

Home Hemodialysis

As an alternative to the in-center treatment, home hemodialysis is encouraged and supported for certain patients. It has been argued that home hemodialysis, once initiated, is less expensive, a more comfortable experience for the patient, and less invasive of the dialysis patient's privacy and lifestyle. The cost of home hemodialysis has been reported to be approximately $5,000–$10,000 per year, as opposed to in-center hemodialysis at a cost of between $20,000–$35,000 per year.[7] Some research has revealed lower morbidity and mortality rates among home hemodialysis patients.[8] In a study of home hemodialysis patients, Atcherson found, however, that clearly, "not all patients or families are capable of coping with the tension and stress of using hemodialysis treatment at home."[9] She discovered, additionally, that the most significant psychological factors, such as anxiety and stress, appeared to be those affecting the dialysis assistant or "partner." Wick and Rye report that a review of the literature revealed that home hemodialysis "provides for better rehabilitation, reduced risk of hepatitis, more independence, freedom to conveniently schedule dialysis and decreased treatment cost."[10] Among the disadvantages were anxieties over machine failure and stress on the patient's assistant or partner. In their own research with home dialysis patients, Wick and Rye found that the most frequently mentioned sources of stress were "needle insertion; machine problems; fear of complications; and blood leaks."[10]

The research of MacElveen et al suggests that patient success with home hemodialysis is notably related to cooperation and understanding among patient, partner, and physician.[11] Emotional adjustment to such hemodialysis was found to be predictable from objective psychological tests administered soon after initiating the program. Fishman and Schneider assert, "For both patients and their relatives, we found that the greater the expression of emotional problems early in the training program, the worse the first year emotional adjustment."[12]

In discussing psychosocial care for the home dialysand and partner, Sramek reports that while the dialysis patient must be encouraged to actively participate in the treatment sessions, the most important aspect of care for the partner is "validifying as normal the feelings of guilt, anger, grief and anxiety."[13]

Intermittent Peritoneal Dialysis (IPD)

IPD was one of the earliest treatment modalities utilized for acute and chronic renal failure. Czaczkes and Kaplan De-Nour report that intermittent peritoneal dialysis was initiated experimentally as early as the 1920s and 1930s. They note that the work of Boen et al[14] in the 1960s, describing an automatic cycling machine, "started the development of the semi-automatic and fully automatic peritoneal machines in use today."[15] Intermittent peritoneal dialysis, which takes place within the abdominal cavity, can be used to treat acute or chronic renal failure. Brundage notes,

> Peritoneal dialysis may be used to sustain life while a patient is being evaluated for chronic hemodialysis or transplantation. It may be used for the initial treatment of severely uremic patients. Peritoneal dialysis with conservative medical management can be offered to many patients who for some reason cannot undergo hemodialysis or renal transplantation.[16]

It has been suggested also that peritoneal dialysis may be the modality of choice for patients wishing to dialyze at home who are unable to carry out home hemodialysis. Intermittent peritoneal dialysis, as performed in medical facilities, may consist of such programs as biweekly infusions consisting of 18–24 exchanges utilizing 1 or 2 liter containers of dialysate, or a dialysate flow of approximately 2.5 liters per hour for approximately 12 hours, 3 times each week. "The one or two liter bottles of dialysate are hung so that the fluid flows by gravity into the peritoneal cavity, where diffusion and osmosis take place."[17] Some of the complications of peritoneal dialysis may include "perforation of the bowel or bladder, rupture of a blood vessel, peritonitis, protein loss, obstruction of the outflow, hypervolemia, and hypovolemia."[18]

Presently peritoneal dialysis, particularly in some of its newer modality regimens—CAPD (continuous ambulatory peritoneal dialysis) and CCPD (continuous peritoneal dialysis)—is being advocated for a number of ESRD patients. To date, CAPD is perhaps the most widely accepted peritoneal procedure for chronic ambulatory patients; and the modality appears to be increasing in availability and accessibility rapidly.

Continuous Ambulatory Peritoneal Dialysis (CAPD)

CAPD was introduced in the mid-1970s as a dialysis alternative to IPD and hemodialysis for the treatment of end-stage renal failure. "With CAPD the patient instills the dialysis solution for four to six hours and exchanges the solution four to five times per day, six or seven days per week."[19] It has recently been suggested that 3 exchanges per day may be adequate for a small percentage of the CAPD population. Identified specifically were those persons with smaller body size, some residual renal function, and capable of totally 1½ liters of drainage.[20] CAPD necessitates the continued presence of dialysate fluid in the patient's peritoneal cavity. "To achieve this a permanent peritoneal catheter is surgically implanted into the patient's abdominal cavity. This catheter is attached to a transfer set which is aseptically connected to a 2 liter plastic bag of dialysate."[21] Fresh dialysate solution is introduced into the peritoneal cavity by gravity and removed similarly, after the required time has elapsed. Patients can usually manage the procedure quite well without assistance. However, they must be cautioned to maintain sterility because of the ever present danger of infection within the peritoneal cavity. CAPD candidates must also be compliant with the treatment regimen, as success of the procedure depends almost exclusively upon self-care.

CAPD provides patients with both physical and psychological freedom. They are independent of the "machine" as well as the continued assistance of family or professional caregivers.[22] Psychological adaptation, however, must be supported; several stages of such adaptation have been identified as the "miracle cure," "reality contact," and "acceptance" phases of adjustment.[23] Finally, it has been suggested that costs of CAPD are considerably less than certain other modalities.[21] The potential for cost-effectiveness of the procedure exists; at present financial reports remain varied.

Several long-term hemodialysis patients in the present study discussed their attitudes toward CAPD. One female patient, who was presently on "self-care" in the hemodialysis unit, stated that she was seriously considering CAPD as an option. She commented, "I'd like to do the treatment myself. I don't want a lot of attention and I don't like people hovering over me." She added, "These units are not as well kept as they ought to be. I think there's a lot of carelessness and shaky technique." Another female patient reported that she had recently discussd the CAPD option with her physician and felt very positive about it, noting, "Anything'll beat going on that machine, let me tell you. You're so tired and drained out when you get off of it, you don't feel like doing anything." Two other female patients, however, expressed negative attitudes toward CAPD, citing body image and the "bag on your stomach" as reasons for their feelings. One patient observed that it had to do with the "cosmetic effects" and her "husband and wife relationship"; the other commented, "I don't like the idea because of the way your body looks. I wouldn't like having that thing on my stomach. I think my sex life would be different. It would really be just uncomfortable with that bag there."

Finally, one study patient who had recently switched from in-center hemodialysis to CAPD spoke positively about the peritoneal procedure. She commented, "It's so nice not to have to go to that place anymore [the hemodialysis unit]. It used to hurt my back so bad just sitting in the chair for all those hours. My back would get to hurtin' so bad I couldn't wait to get off." Of CAPD, she reported, "This new one is not too bad except sometimes I don't feel like foolin' with those bags 'n' all. But it's okay. My stomach feels kind of swelled up at first though, but then you get adjusted to it."

Continuous Cyclic Peritoneal Dialysis (CCPD)

Advances in technology have brought about the introduction of a home peritoneal dialysis modality labeled CCPD, which conceivably can be carried out at night while the patient sleeps. Generally, CCPD involves 3 exchanges of dialysate fluid in the abdominal cavity per session and takes approximately 9 hours to complete. Automatic dialysis delivery systems "warm the dialysate and regulate the inflow and outflow for the whole treatment, eliminating the need to repeatedly enter the system to connect and disconnect containers and drainage bags."[24] Dialysis patients can learn to operate this machine quite easily and may dialyze alone without anxiety. One salesperson for a company manufacturing the CCPD equipment noted that a home patient coordinator is available around the clock to answer questions or give advice, and added that needed machine repairs could be made within 24 hours. Two specific advantages of home peritoneal dialysis are that it is less expensive than hemodialysis and requires a relatively brief period of training.[25] A summary of the advantages of CCPD over CAPD was suggested in 1981 by Diaz-Buxo et al:

Continuous cyclic peritoneal dialysis (CCPD) incorporates the physiological advantages of CAPD, eliminates the frequent exchanges of dialysate during the day, and provides automated nocturnal exchanges with the addition of a cycler. The rationale for this design is to provide more convenience and potentially reduce the incidence of peritonitis.[26]

Several new initiates to the CCPD procedure were interviewed about their early adaptation to the modality of treatment. One female patient reported that although she had been assured that she could manage the procedure herself, she wanted her husband to participate in the training sessions, commenting, "It will be good if he knows what to do, too, if something goes wrong. There is this little beeper; you hear it in your sleep, believe me; but I still don't sleep too good. It's a worry." A male CCPD patient spoke very positively about the procedure; his wife, however, expressed some ambivalence toward the automatic dialyzer. She reported that she had now moved out of the bedroom that the couple formerly shared, commenting, "I told him that that machine is now your mother, your wife, your friend, and your lover!" She added,

I was sleeping with him at first because I had to be there. I had to be right there. Then one day I got up to eat and heard that thing "beepin" down the hall and I figured that why should I sit up half the night fussing. At first I was scared and tense and I had to be right there. But now I go on to my room. See that's his room, this is my room—and that's the way we do it now.

Another male CCPD patient reported that he was not pleased with the procedure, specifically because he felt very uncomfortable carrying around extra fluid in his abdomen all day. He stated, "I can't carry all that 2000 [2 liters] of fluid in me because my body absorbs too much." He added, "Some people's systems are different." The patient's wife, who was working full-time, commented that she had recently been having severe headaches and attributed it to the stress of coping with the nightly dialysis procedure. She stated,

> I don't sleep at night worrying about him on this machine. He's got the whole 2000 [2 liters] in and out of him overnight and I'm afraid that that last 2000 is going to go in and out before he can shut the machine off; I'm afraid it's gonna drain out of him. And he's not gonna be able to hold it in him. And I don't sleep sound; every once in a while, I turn to him and say, "Is that machine all right? Are the bottles running okay?" I mean, this has been going on for about two months and I just don't sleep. I don't get my right rest.

Kidney Transplantation

For approximately the past two decades renal transplantation has been a clinically acceptable option for the treatment of end-stage renal failure. "Modern clinical transplantation began in Boston in 1954 when Murray and others performed a transplant between identical twins."[27] This transplant was based on much preliminary investigative work by Dr. David Hume.[28] It has been noted, however, that there was little chance of long-term graft survival until the initiation of immunosuppressive therapy, utilizing such drugs as Imuran and Prednisone. By the early 1960s the ability of immunosuppressive drugs to delay graft rejection was established,[29] and "in 1962 the first successful series of cadaver renal transplants were performed, primarily by Dr. Hume in Richmond, Dr. Thomas E. Starzl in Denver, and Drs. Joseph Murray and John P. Merrill in Boston."[31] Kagan asserts that "with the identification of histocompatability antigens and the use of immunosuppressive drugs, the overall long-term success of the graft presently approaches 90 percent when siblings are used as kidney donors"[30] Graft survival rates are notably lower in cadaver transplants. As Fox and Swazey noted in 1978, "General recognition of the impermanence of non-related cadaver kidneys compared with those transplanted from living related donors has been firmly, if somewhat regretfully established."[31] While it is reported that "30,000 persons" have received kidney transplants for their renal failure,[32] investigation and experimentation related to the overall renal transplantation regimen continue. The future remains uncertain—yet the picture seems to have brightened recently with the advent of a new immunosuppressive,

Cyclosporine, providing surgical teams with a guardedly optimistic view of transplant success to come. The result has been an almost zealous reaffirmation of the positive benefits of the transplant procedure and a heightened awareness of the need for cadaver and living related donor kidneys.

Maintenance hemodialysis patients in the present longitudinal study were asked to articulate their attitude toward renal transplantation, and some attempt was made to determine whether any relationship existed between compliance with the dialysis regimen and opinions on a possible future transplant. Analysis of T1 data on the 126 original study patients revealed that 70 (55.6 percent) of the respondents were negative toward renal transplantation for themselves, 15 (11.9 percent) were uncertain, 19 (15.0 percent) felt positive about such surgical intervention, and 22 (17.5 percent) held strongly positive attitudes toward this modality of treatment. It was found that those persons who held negative opinions about transplantation demonstrated the lowest degree of compliance behavior and those who held positive opinions, the highest. The reasons for dialysis patients' attitudes about kidney transplantation are generally complex, and interpretation in terms of acceptance or nonacceptance of a suggested surgical procedure tends toward oversimplification, as the following statements illustrate:

1. *Female, 58 years, married, on dialysis one year:* "I don't want a transplant. I feel that when my time is up, God will take me. I don't want any foreign things in my body."
2. *Male, 35 years, married, on dialysis 2 years:* "I'm worried about the reactions [of transplantation] as a result. I would have to have a cadaver. I worry about the use of all those drugs—the operation is not yet perfected enough."
3. *Male, 49 years, married, on dialysis over 3 years:* "At first, I was apprehensive about the transplant. I saw my roommate have some problems. Then I saw someone who had one eating pie and all kinds of things—this made up my mind to want it."
4. *Female, 56 years, widow, on dialysis one year:* "I don't like the idea [of transplant]—seems like I just want what I was born with—not something that belonged to somebody else."
5. *Male, 33 years, married, on dialysis 2 years:* "I don't know about a transplant— I do and I don't [want one] because I'm doing well on dialysis. Maybe I'll wait. What got me was seeing some that didn't work—I'm afraid."
6. *Female, 33 years, married, on dialysis 4 years:* "At present I don't want one [a transplant] because of the uncertainties related and the experimental attitudes of the doctors treating patients with the disease and the kidney—some try to keep the kidney in, even if it's rejecting."

(It should be pointed out that recent reports indicate that decreased morbidity and mortality rates for kidney transplantation are related to patient survival rather than graft survival.)

Several of the long-term dialysis patients defended their negative attitudes toward transplantation by discussing others' experiences. One younger male patient stated, "I don't want one [a transplant] right now unless I could get a real good match. I asked for only a 3 or 4 antigen match and they asked me to go down to a 1 or 2, but I said, 'No.' " He continued,

Do you remember Mr. A.? He was here [in the dialysis unit] a couple of years ago—a 40-year-old guy who got a living related from his mother. He rejected in a few months and about 8 months later, just when he was getting himself back together he got another living related from his sister and then he ended up dying a few days later from a septicemia.

Another male patient shared his experiences: "My brother had a transplant and it rejected and he died. I've seen a lot of patients out here who got a transplant and then they died. Sometimes your heart won't take it."

Finally, a few long-term patients reported that they either did not wish or were not able to have transplants for family donor-related reasons. One woman who had been on dialysis 9 years summed it up this way: "I have thought about it [transplant] and my daughter, she has offered me her kidney, but I was afraid to take it in case one day she lost her one kidney and would need the other. I don't care much for them cuttin' on her body either." Another female patient recalled that she had not accepted her brother's offer of a kidney, adding, "I don't want to be responsible for his death." A 52-year-old male patient reported that his family didn't talk much about transplant. He went on, "My sister mentioned giving me a kidney about a year ago—but didn't repeat it. I feel like I can't ask. I just have to wait and see if it's mentioned again." Another male patient who had rejected two cadaver transplants described his experience this way: "My sister, she don't like to get in contact with me because when I had the first transplant a lot of her children wanted to get involved. She didn't like that too much though. She was trying to protect her kids. She didn't want them to take a chance of givin' me a kidney."

TRANSPLANT REJECTION: RETURN TO THE DIALYSIS UNIT*

Reentry into the dialysis unit is beginning to occur more frequently with the increasing number of kidney transplantation procedures and the patient's risk of possible rejection. While some patients appear to cope fairly well with the loss of a transplanted kidney and with the subsequent neces-

---

*This section was originally published in O'Brien ME: Return to the hemodialysis unit—coping with kidney transplant rejection. Am J Nurs (in press). Permission to quote obtained from the publisher.

sitated return to dialysis, others experience anger, grief, and even severe depression in dealing with the adjustment.

For the chronic renal failure patient there are at present only 2 choices: life-long dialysis treatments or renal transplantation.[33] Alternatives are generally weighed carefully by patients in an attempt to determine whether transplantation with all of its uncertainties is nevertheless of greater benefit than continued dependency upon a dialysis modality and the related aspects of that treatment regimen.

Hemodialysis, peritoneal dialysis, and renal transplant patients often fear "living the unsatisfactory life of the chronically-ill and handicapped person,"[34] but for many renal patients, transplantation appears as a "savior,"[35] bringing with it many aspects of a renewed life. Such patients realize that a successful transplant may free them from the long and tedious hours on a dialysis machine or the daily coping with an invasive peritoneal procedure.

Unfortunately, a patient electing transplantation may not seriously consider the very real possibilities of graft rejection and subsequent return to the dialysis treatment regimen. The rejection phenomenon "accounts for a high incidence of graft loss within the first three months after surgery,"[36] and the risk of possible rejection is always present. The 3 major types of graft rejection are *hyperacute rejection,* which occurs "during the surgical procedure when blood vessels are anastamosed"[37] or very shortly thereafter; *acute rejection,* occurring "usually with the first ten days"[37] after transplantation or possibly recurring within the first few months; and *chronic* (or late) *rejection,* which "becomes apparent months or years after the transplant and gradually reduces the kidney function.[37] It has been found that rejection episodes "occur in approximately 85 percent of all transplanted patients,"[38] and although acute rejection is often reversible, there are occasions when it cannot be satisfactorily reversed "due to the immunological response of the host."[38]

When the rejection episode is irreversible, the graft must be removed and the dialysis treatment procedure reinstituted. Often this loss of the transplanted kidney results in great disappointment on the part of both the dialysis patient and his or her family. If the graft has been provided by a sibling or close relative (i.e., a living, related donor) uniquely serious adjustment problems may occur. Fox and Swazey note that "the recipient who does not retain an organ has been known to experience guilt because he feels that he has willfully rejected both the organ and its donor."[39]

In addition to the psychosocial adjustment that must be made by the transplant-rejected dialysis patient, often new or more severe physical problems are encountered relative to the trauma of the surgery. Patients sometimes complain of pain or discomfort in the area of the surgical site, experience loss of appetite and energy, and occasionally have to face

complicated access problems due to graft or fistula deterioration during the
surgical and postsurgical periods.

The Transplant Reject Patient

In the present study of long-term hemodialysis patients,[40] 7 of the original 126 patients were found to have undergone unsuccessful kidney transplantation surgery and had been forced to return to dialysis. Of the patients who had rejected their transplants, 6 were male and 1 female; ages ranged from 37 to 51 years; educational achievement ranked from the level of "grade-school only" to "college degree" (baccalaureate); length of time on dialysis prior to transplant ranged from 1 to 4 years. All of the patients had had cadaver transplants, one having had 2 cadaver transplants approximately 20 months apart. Hospitalization periods for the transplant surgeries ranged from 2 weeks (for a patient experiencing hyperacute rejection) to 4 months. The types of rejection phenomenon experienced included one case of hyperacute rejection (occurring on the operating table); 4 cases of acute rejection that could not be successfully reversed (patients' hospitalizations ranged from approximately 3 weeks to 6 months); and 2 cases of chronic or late rejection in which the transplanted kidneys were retained for 1 year and 1 year and 4 months, respectively.

In interviews with the study patients after the transplant rejection and subsequent return to the hemodialysis unit had occurred, the majority reported that loss of the kidney had been very difficult but that they were generally happy to be alive. All noted that both they and their families had been very positive initially toward the idea of transplantation, and one patient who had experienced late rejection stated that his family had "really enjoyed" his year of transplantation. He suggested that his life was very different after the surgery: that he had more freedom, did not feel tied to a machine, and could do what he wanted and eat what he wanted. This patient admitted that he had "felt pretty badly about the rejection because it was a slow [chronic] rejection," but that "he knew it might come." He also noted that he had accepted going back on dialysis fairly well because "you don't expect things to last forever," but added, "they're looking for another one [kidney] now. I'm ready and I'll keep trying."

Four of the patients stated explicitly that their reason for undergoing kidney transplantation was to "get off the machine," and two asserted that they expected the transplant to make "life better" and questioned the quality of their lives on hemodialysis. One patient, a 51-year-old male who had been on dialysis for approximately 4 years prior to transplant and who had experienced an acute irreversible rejection episode only 3 weeks postsurgery, appeared quite depressed. He reported that it had been "a big disappointment when it [the kidney] rejected," and stated, "I had my [own] kidneys taken out and now I am completely without any kidneys." He added that many physical problems had occurred since the rejection, and he speculated, "If I hadn't had the transplant, I would be working now." This patient noted that he really hadn't had his transplanted kidney long enough for it to begin to work, so he didn't really know whether transplantation had made any difference in comparison with his life on dialysis. He stated that he would not consider another transplant for some time.

Two other patients admitted to experiencing great difficulty in going back on dialysis, both of whom had retained a transplanted kidney for only 3 to 4 months. Both stated at the time of their interviews that they were then unwilling to consider another transplant and neither looked forward to such a possibility in the future. One patient commented that "he wished to get his body back together" before attempting another transplant, but suggested that he might try again after some time had passed. The other 4 patients reported that they would consider kidney transplantation again as soon as a compatible donor kidney could be found.

Interestingly, the two patients who had retained their transplanted kidneys for over a year reported fewer complaints in regard to the rejection experience than the other study reject patients. Although they expressed sorrow over the loss, these patients said that it had not been "too bad" to return to hemodialysis and both were anxious to receive another kidney as soon as possible. One might speculate therefore that a longer time of transplant retention leads to a more positive attitude toward future transplantation and thus modifies to some degree the stress of return to hemodialysis. Those patients who had retained transplanted kidneys for shorter periods (approximately 3 weeks to 6 months) tended to find the rapid return to dialysis more difficult and disappointing, admitted to more physical complaints on resumption of hemodialysis, and generally held much more negative attitudes toward the possibility of future kidney transplants. The variables of age, sex, length of time on dialysis prior to transplantation, and the presence of family support systems did not appear influential in the patients' adjustment to transplant rejection and return to dialysis, nor to their attitudes toward future transplantation.

Chronic Dialysis: Toward a Treatment Prototype

"Chronic dialysis," whether carried out as hemodialysis or one of the peritoneal modalities, may serve as a treatment prototype for future chronically ill patients dependent upon sophisticated technology. As such, the dialysis experience would have important ethical and practical implications for future research and development in the general area of medical care technology. Researchers and other scholars will need to look at the juxtaposition of technology and humanity and seek to understand and publicly articulate the long-term implications of such health care modalities, not only for the individual but for society as a whole. As a political community, American society has formally committed itself, through the development of legislation, to continuing support of patients with end-stage renal disease. One must look, however, at the needs of the rest of the community as the financial resources targeted for health care balloon and are more and more widely perceived as excessively burdensome. More costly medical problems must also be anticipated for dialysis and other patients as the American population continues to age. As these issues are examined, end-stage renal disease and the chronic dialysis experience may serve to provide society with directions for future support and intervention in other life-threatening illness conditions.

THE FUTURE OF DIALYSIS: A RETURN TO SCARCE RESOURCES?

Prior to 1960, approximately "55,000 American patients died each year" from irreversible end-stage renal failure.[41] In the early days of hemodialysis history in this country, notably the 1960s, however, patients were presented with a somewhat mixed blessing in the medical community's acceptance of the use of the "artificial kidney" as a treatment modality for end-stage renal disease. Although the artificial kidney machine was being made available at a number of medical centers across the United States, the cost of treatment was such that few individual patients could maintain the procedure on a continuing basis. Initially in-center dialysis twice a week cost "approximately $20,000 per patient per year."[42]

While public discussion of the hemodialysis treatment and its implications for ESRD patients in the media generated some funding from both public and private sources, the resources remained nevertheless scarce. Thus, a crucial question arose as to who would receive the life-saving treatment. As noted previously, hospital or "selection committees" affiliated with medical centers, sometimes referred to ironically as "God Squads," were assigned as gatekeepers of the then scarce resources.

With the 1972 advent of Public Law 92-603, however, providing federal coverage of treatment costs for all ESRD patients, the picture brightened notably. Selection committees became obsolete and the hemodialysis procedure became available and accessible to all renal failure patients. As Fox observed, "What seems to have happened, largely as a result of Public Law 92-603, is that virtually no one with end-stage renal disease is being excluded from chronic dialysis, regardless of what contra-indications might exist."[43] Obviously this condition of now "abundant resources" has led to the raising of new questions about who should be dialyzed, relative to quality of life. These questions remain. Related to these selection issues is an increasing concern in both public and private sectors about the phenomenally escalating cost of the federally funded ESRD program and the fear of return to an era of scarce resources if adequate cost containment measures are not developed and implemented.

The ESRD program has grown far beyond the expectations of Congress, which legislated the support, and of the physicians who proposed it. It has been noted that "more patients immediately became candidates for long-term hemodialysis once public funding became available, and they have lived longer than originally anticipated."[44] It had also been anticipated that kidney transplantation and home dialysis would significantly alleviate the initial financial burden incurred with in-center hemodialysis. This has not been the case. Due in part to the lack of available donor kidneys, transplantation has not escalated as rapidly as expected; and home hemodialysis, which once initiated is notably less expensive than in-center treatment, currently is used by only about 17 percent of the national dialysis patient population.[45]

The ESRD program in 1982 serviced approximately 68,000 Americans "at a Medicare cost of $1.2 billion dollars, an amount reportedly representing 3.8 percent of the total Medicare benefit payment.[46] Roberts at al note that these funds are spent on only about .016 percent of the population and suggest that if dialysis populations remain consistent, then new patients entering the system must be directed to other modalities if any cost containment is to be maintained.[47]

It is speculated that by 1985 the ESRD program could cost Medicare $2.68 billion to serve 83,700 patients.[48] Thus, a concern necessarily surfaces as to whether the country, specifically the federal government, will continue to fund the program at its present availability and accessibility levels. Questions may arise regarding the priority of ESRD patients' needs over those of patients faced with other life-threatening illness conditions. It may be necessary to examine again the issues of prevention versus treatment and research to seek causes and cures versus continued maintenance therapy. Finally, the medical community may be forced to face again a condition of scarce resources, with its necessarily related dilemma of who shall live and who shall die. The future of the ESRD program cannot be clearly predicted, but unless adequate cost containment measures are soon implemented, continued federal funding at the present level is in serious jeopardy.

THE COURAGE TO SURVIVE

The research presented in this book is dedicated to all those patients and their families who have had the courage to live, to love, and to survive while coping with the many and diverse life modifications necessitated by dialytic therapy. It is dedicated also to the caregivers who choose to give of themselves in a chronic illness involvement, knowing that the patient–practitioner relationships upon which they embark may well be cast in the context of " 'til death do us part."

Although survival with end-stage renal disease is susceptible to romantic portrayal, it is not a "normal" situation and the life on chronic dialysis is not a "normal" life. ESRD is not a nice disease. Frequently, however, the gravity of the condition appears to be denied by caregivers, family members, and even the patients themselves in order to facilitate the business of getting on with the tasks of daily living. The life-threatening illness and its complex treatment modality are routinized, with both positive and negative effects for patients. The positive outcome of such acceptance and routinization is the patient's ability to carry on with his or her life in as satisfactory a manner as possible, given individual physical limitations. Certain negative consequences, however, may result from an overzealous attempt to normalize the situation, prohibiting the patient from expressing the healthy emotional reactions of anger, anxiety, and frustration at the truly overwhelming life restrictions and stresses imposed by the illness condition. The sufferings of the patients are many, both physical and psychosocial—pain, fatigue, anxiety, depression. Of all, however, perhaps the most difficult correlates of the chronic dialysis patient status are loneliness and uncertainty.

In the beginning, especially, hemodialysis patients are surrounded by family, friends, and caregivers who attempt to help them cope with the newly initiated treatment and its life restrictions. For some patients, such support remains constant over time, but for others it gradually diminishes, as the novelty of the condition and its treatment modality diminishes. For all, the loneliness of being a hemodialysis patient remains constant. Alone and uncertain of the future, one must thrice weekly surrender one's physical being to the magnificent yet fearful technology that both threatens and sustains life. Alone and uncertain, the dialysis patients submit their will—desires, hopes, goals—to the medical team who advise and guide a "career" in chronic illness. Alone and uncertain, they face the possible disruption of family life and social interaction, the modification or elimination of role responsibilities, and the limitation of physical functioning. The chronic hemodialysis patients face the uncertainty of their future, living daily, perhaps hourly, with the knowledge of their total dependence upon a machine for the continuance of life. Many persons may be present to the dialysis patients—many may sympathize, many may empathize—but none is able to cross over, to stand with the patients in that place of illness where they are alone, and truly understand.

In sum, long-term dialysis patients must be possessed of "the courage to survive"—to survive in pain, in fatigue, in anxiety, in depression and, most especially, to survive in loneliness and uncertainty. For it is when one feels truly alone, when one feels truly uncertain, that courage is most needed, is most essential. For the chronic hemodialysis patients, courage is not only a necessity for survival, courage is survival—courage is life!

REFERENCES

1. Duff R, Hollingshead A: Sickness and Society. New York, Harper and Row, 1968
2. Gutman RA, Stead WW, Robinson RR: Physical activity and employment status of patients on maintenance dialysis. N Engl J Med 304:309–313, 1981, p 311
3. Relman AS, Rennie D: Treatment of end-stage renal disease: Free but not equal. N Engl J Med 303:996–998, 1980
4. Lowrie EG, Laird NM, Parker TG, Sargent JA: Effect of the hemodialysis prescription on patient morbidity. N Engl J Med 305:1176–1181, 1982, p 1176
5. Czaczkes JW, Kaplan De-Nour A: Chronic Hemodialysis as a Way of Life. New York, Brunner/Mazel, 1978, p 43
6. Kjellstrand CM, Avram MM, Blagg CR, Friedman EA, Salvatierra O, Simmons RL, Williams GM, Terasaki P: Cadaver transplantation versus hemodialysis. Trans Am Soc Artif Intern Organs 26:611–620, 1980, p 613
7. Sramek JR: Psychosocial care for the home dialysand and partner. Nephrology Nurse 3:37–42, 1981, p 37
8. Kolodner L, McCuan E, Levenson J: Screening and supportive techniques for home dialysis in the treatment of renal failure. J Am Geriatr Soc 24:32–35, 1976
9. Atcherson E: The quality of life: A study of hemodialysis patients. Health Soc Work 3:55–69, 1979, p 64
10. Wick GS, Rye DZ: Home dialysis, is it worth it? J AANNT 4 (supple):6–13, 1977, pp 7, 9
11. MacElveen PM, Hoover PM, Alstander RA: Patient outcome success related to cooperation among patient, partner and physician. J AANNT, 4 (supple):148–156, 1977
12. Fishman DB, Schneider CJ: Predicting emotional adjustment in home dialysis patients and their relatives. J Chron Dis 25:99–109, 1972, p 107
13. Sramek JR: Psycho-social care for the dialysand and partner. Nephrology Nurse 3:37–42, 1981, p 41
14. Boen ST, Mion CM, Curtis FK, Shilipetar G: Periodic peritoneal dialysis using the repeated puncture technique and an automatic cycling machine. Trans Am Soc Artif Intern Organs 10:409–415, 1964
15. Czaczkes JW, Kaplan De-Nour A: Chronic Hemodialysis as a Way of Life. New York, Brunner/Mazel, 1978, p 6

16. Brundage DJ: Nursing Management of Renal Problems. St. Louis, C.V. Mosby, 1976, p 111

17. Brinkley LS: Maintenance peritoneal dialysis, in Lancaster L (Ed): The Patient with End-Stage Renal Disease. New York, Wiley, 1979, pp 209–238, p 209

18. Kagen LW: Renal Disease. New York, McGraw-Hill, 1979, p 147

19. Weinman EJ, Senekjian HO, Knight TF, Lacke CE: Status report on continuous ambulatory peritoneal dialysis in end-stage renal disease. Arch Intern Med 149:1422, 1980, p 1422

20. Simon P, Moncrief JW, Pyle K: CAPD: Are three exchanges per day adequate? J AANNT, 9:39–43, 1982, p 43

21. Arenz R: Do-it-yourself dialysis. Registered Nurse 7:57–60, 1981, pp 57–58

22. Ainge RM: Continuous ambulatory peritoneal dialysis. Nursing Times 9:1036–1038, 1981, p 1038

23. Mehall DL, De Young K, DeYoung M: The psychological adjustment of a CAPD patient. J AANNT, 8:23–24, 1981, p 23

24. Denniston DJ, Burns KT: Home peritoneal dialysis. Am J Nurs 80:2022–2025, 1980, p 2022

25. Irwin BC: Now—peritoneal dialysis for chronic patients too. Registered Nurse 44,6:49–52, 1981, p 50

26. Diaz-Buxo JA, Walker PJ, Chandler JT, Farmer CD, Holt KL: Advances in peritoneal dialysis—continuous cyclic peritoneal dialysis. Contemporary Dialysis 11:23–26, 1981, p 23

27. Brundage DJ: Nursing Management of Renal Problems. St. Louis, C. V. Mosby, 1976, p 145

28. Dunphy JE: The story of organ transplantation. Hastings Law Journal 21:1–6, 1969

29. Lee HM, Thomas FT: Cadaver renal transplantation. Dialysis and Transplantation 6:30–37, 1975, p 30

30. Kagan LW: Renal Disease. St. Louis, McGraw Hill, 1979, p 215

31. Fox RC, Swazey JP: The Courage to Fail (2nd ed). Chicago, The University of Chicago Press, 1978, p 307

32. Taylor JH, Hopper SA, Pierce P: The patient receiving a renal transplant, in Lancaster L (Ed): The Patient with End-Stage Renal Disease. New York, Wiley, 1979, pp 239–292, p 242

33. Kobrzycki P: Renal transplant complication. Am J Nurs 77:641–643, 1977, p 641

34. Beard BH: Fear of death and fear of life, the dilemma in chronic renal failure, hemodialysis and kidney transplantation. Arch Gen Psychiatry 21:373–380, 1969, p 380

35. Bailey GL: Psychosocial aspects of hemodialysis, in Bailey GL (Ed): Hemodialysis, Principles and Practice. New York, Academic Press, 1972, pp 424–439, p 431

36. Sachs BL: Renal transplantation, a nursing perspective. Flushing, N.Y., Medical Examination Publishing Co., 1977, p 77

37. Gutch CF, Stoner MH: Review of Hemodialysis for Nurses and Dialysis Personnel. St. Louis. CV Mosby, 1975, p 144

38. Kobrzycki P: Renal transplant complications. Am J Nurs 77:641–643, 1977, p 642

39. Fox RC, Swazey JP: The Courage to Fail (2nd ed). Chicago, The University of Chicago Press, 1978, p 30

40. O'Brien ME: Effective social environment and hemodialysis adaptation: A panel analysis. J Health Soc Behav 21:360–370, 1980

41. Ellison DL: The Bio-medical Fix. Westport, Conn., Greenwood Press, 1978, p 102

42. Fox RC, Swazey JP: The Courage to Fail (2nd ed). Chicago, The University of Chicago Press, 1978, p 208

43. Fox RC: Exclusion from dialysis: A sociological and legal perspective. Kidney International 19:739–751, 1981, p 744

44. Greenspan RE: The high price of federally regulated hemodialysis. JAMA 246:1909–1911, 1981, p 1910

45. Robinson D: Kidney dialysis: A taxpayer's nightmare. Readers Digest, October 1982, pp 149–152, p 152

46. Capelli JP: Testimony of the Catholic Health Association of the United States, on proposed end-stage renal disease regulations, before the Subcommittee on Oversight, Committee on Ways and Means, United States House of Representatives, April 22, 1982, p 1

47. Roberts SD, Maxwell DR, Gross TL: Cost-effective care of end-stage renal disease: A billion dollar question. Ann Intern Med 92:243–248, 1980, p 248

48. Greenspan RE: The high price of federally regulated hemodialysis. JAMA 246:1909–1911, 1981, p 1909

─────Appendix─────
The Research Program: Evolution and Progress

STUDY PURPOSE

The initial study phase, T1, examined quantitatively the association between the support of significant others and selected aspects of social and social–psychological functioning, conceptualized to include interactional behavior, quality of interaction, sick role behavior, secondary gain, and alienation. Also investigated were the variables of sexual functioning, attitude toward transplantation, and the import of religious faith in adjusting to ESRD and maintenance hemodialysis. During the second and third phases of the study (T2 and T3), the relationship between significant others' support and patients' social functioning was re-examined in an attempt to identify changes over time in correlations. Significant variations over time in individual study variables were explored as well. The third phase (T3) sought also to generate qualitative data directly from each patient and from family, friends, and caregivers in an attempt to describe dominant themes in the life career of the chronic dialysis patient.

PROGRAM PHASES

Phase I: (T1—1974–1975).   126 adult hemodialysis patients were interviewed utilizing a structured interview schedule.
Phase II: (T2—1977–1978).   63 of the original 126 patients were identified as a panel group and interviewed utilizing the same interview schedule.
Phase III: (T3—1980–1981).   33 of the 63-member patient panel identified at T2 were again interviewed utilizing the previously constructed interview schedule. Unstructured, focused interviewing was also carried out with the 33 member patient panel. Focused interviewing was carried out with 26 significant others (family members/friends) identified by the patient panel. Focused interviews were conducted

with 45 dialysis unit caregivers—head nurses, staff nurses, therapists, social workers, physicians, technicians, and clerks. Direct, nonparticipant observation was carried out in 3 large metropolitan hemodialysis units.

Phase IV: (T4—1982–1983). A clinical nursing intervention study based upon previous research was initiated with a group of new maintenance dialysis patients.

THEORETICAL FRAMEWORK

The theoretical framework undergirding the initial study phase was that of symbolic interactionism, primarily as developed by George Herbert Mead. Focus was placed upon Mead's assertion that one's behavior may only be understood in terms of social group memberships and that most human actions are predicated upon one's perceptions of the attitudes and expectations of significant others within one's social system.[1]

Mead employed the term "significant other" to refer to those persons whose behaviors, attitudes, and expectations were considered important and/or influential by an actor in determining his or her own ongoing behavior. For the present research the term 'significant other' was applied to family, friends and those contractually, yet importantly related to the hemodialysis patient, such as physicians, nurses and therapists involved with the respondent and his or her medical regimen.

STUDY VARIABLES

Effective Social Environment

Significant others' support was identified under the label *effective social environment.* As described by Duff and Hollingshead, effective social environment is "a social setting encompassed by family; members of kinship groups; work associates; friends and others who are meaningful to the ill person."[2] For the present research, effective social environment was conceptually divided into 2 systems:

1. *Primary system*—identified in terms of affective relationships, i.e., family and/ or close friends.
2. *Secondary system*—identified in terms of instrumental relationships, i.e., health caregivers.

Social and Social–Psychological Functioning

Social and social–psychological functioning was described as consisting of selected variables denoting patient attitudes and behaviors such as social activities and interactions with significant others in institutional and community contexts. These variables include interactional behavior, quality of interaction, sick role behavior, secondary gain, and alienation.

1. *Interactional behavior*—implying social and social–psychological activities and interactions (e.g., in family-kinship, friendship, religious, recreational and work-related contexts).
2. *Quality of interaction*—dealing with the patient's self-reported satisfaction in interactions with significant others, including such factors as ease or difficulty in establishing new relationships and facility in getting along with other members of the family.
3. *Sick role behavior*—conceptualized in terms of the patient's health/illness behavior as this is potentially modified or changed by his or her disability. Sick role behavior is defined theoretically as that behavior deemed appropriate under existing societal norms for an ill person, relative to the nature and severity of his or her condition.

Parsons' conceptualization presents the sick role as consisting of "a set of institutionalized expectations and the corresponding sentiments and sanctions."[3] These expectations include temporary exemption from social role responsibilities, exemption from responsibility for illness or care, obligation to want to get well, and obligation to seek and to cooperate with technically competent help.

While it is true that sick role behavior as thus conceptualized by Parsons must be modified or respecified in order to apply to chronic illness, certain precedents for such application have been established.[4] In the present context the conceptualization of sick role behavior was restricted to "compliance with the treatment regimen," thus focusing on only two of Parsons' expectations, i.e., the obligation to want to have major symptoms alleviated and, to that end, the seeking of and cooperating with technically competent help.

4. *Secondary gain*—employed in the positive sense as the renal patient finding any "positive" or satisfying elements in his or her modified lifestyle resulting from the necessity for continued hemodialysis treatment (e.g., frequency of satisfaction-producing experiences resulting from illness and its regimen).
5. *Alienation*—the concept of alienation, or anomie, as employed in contemporary sociology, was intended to reference what is most commonly explained as isolation and "normlessness."[5]

OPERATIONALIZATION OF VARIABLES

Effective Social Environment

Significant others' support systems were operationalized by means of 2 scales measuring (1) patients' perceptions of the expectation of primary system members, and (2) patients' perceptions of the expectations of secondary system members.

Social and Social Psychological Functioning

Operationalization of the concept of social and social psychological functioning proceeded from questions dealing with the patient's attitudes, social activities, and interactions as modified by his or her particular disability. Data deriving from these questions supported the generation of 5 Likert-type scales: Interactional Behavior (IBS), Quality of Interaction (QIS), Sick Role Behavior–Compliance Scale (CS), Secondary Gain (SGS), and Alienation (AS).

The Interactional Behavior Scale (IBS)

The IBS provides a measure of the patient's relative social involvement/ withdrawal (behavior) in such areas as family, work, recreation, and church and club attendance. Activities dealing specifically with family-related matters are measured in terms of items such as care of the family accomplished by the patient, contribution to household chores accomplished by the patient, carrying out of usual family responsibilities (such as managing the budget, advising or disciplining the children), participating in the activities and experiences of other family members, and frequency of entertaining relatives and friends in the home. Certain items in the IBS had been adapted from the "Disability Self-Conception Inventory" utilized by M. Bridgford in research on the social rehabilitation of the chronic hemodialysis patient.[6]

At the time of initial measurement, detailed data were also collected on social activities "before the onset of renal failure," in order to control for prior levels of social functioning. For each item in the IBS, respondents were asked to provide a retrospective comment evaluating their former degree of activitiy relative to the particular aspect of behavior dealt with in the question. In the main, respondents were found to have been socially active and responsible before the onset of chronic illness.

Quality of Interactional Behavior (QIS)

QIS was operationalized through the construction of a scale consisting of several items measuring the ease or difficulty of the chronic renal patients' social functioning, such as their facility in making new friends or acquaintances and degree of ease of functioning in ongoing family relationships.

Sick Role Behavior

Sick role behavior was operationalized in terms of an 8-item hemodialysis regimen compliance scale eliciting the patients' self-report regarding such factors as attendance at scheduled treatment procedures, adherence to fluid and dietary restrictions, taking of prescribed medications, and general faithfulness to the physician's orders relating to such activities as rest and exercise. Reliability of patients' self-report was confirmed by obtaining a general evaluation of their compliance from appropriate dialysis center personnel.

Secondary Gain

As a subcomponent of sick role behavior, the concept of 'secondary gain' was measured by means of a scale that sought to elicit from the patients indications of any "positive" or satisfying development serendipitously arising from the modified lifestyle necessitated by the hemodialysis regimen.

Alienation

Alienation, or anomie, was measured by use of the standardized "Dean Alienation Scale," which measures alienation in terms of three subcomponents: powerlessness, normlessness, and social isolation.[7]

QUANTITATIVE METHOD: PHASE I, II, AND III

Sample

Initially (1974–1975), a purposive sample was drawn from 3 hemodialysis centers located within an urban university medical center, a military facility, and a city general hospital, respectively, as these were considered representative of various types of major medical institutions in the United States.[8] All patients who met the research criteria were approached by the interviewer, the criteria specifying that sample subjects be between the ages of 21 and 70 years, have been receiving dialysis treatment for at least the preceding 6 months, and have no other serious medical or psychiatric complications (other than ESRD).

Procedure

Study interviews were conducted initially over a 6-month period, and 3 years later, over a 3-month period. Then, 6 years later, extended study interviews continued for approximately 9 months. Appropriate informed content procedures were carried out. Patients were seen both in "centers" and in homes.

Instrumentation

The structured interview schedule developed in 1974, which was employed in the first 3 study phases, consisted of a total of 97 questions eliciting, as well as sociodemographic data, subjects' responses relative to their perceptions of the expectations of significant other and patient behaviors in terms of the selected aspects of social and social–psychological functioning. To control for possible intervening variables related to the history of maturation of the panel group, a number of additional questions were added to the interview schedule at T2. These items served to document the occurrence of social and physical problems during the 3-year intervals, as well as to identify incidents related to difficulties with access or with the dialysis treatment procedure itself.* Similar control items were included in the interview schedule employed at T3.

Data Analysis

Scale-to-scale correlation of major variables were computed utilizing Spearman's rank order, Kendall's Tau and Pearson's product moment procedures.[†] The 3 coefficient sets thus produced—i.e., rho, tau and r—suggested analogous patterns of association between the study variables (i.e. concerning fact, strength, and nature of the relationship). A 2-tailed T-test was employed to determine significance of changes over time on the variables under study. Mean responses on individual scales were evaluated relative to selected demographic and associated variables through computation of a one-way analysis of variance (F-ratio).

QUALITATIVE METHODS: PHASE III

Sample

Study subjects involved in the qualitative aspect of the rescarch initiated at T3 consisted of the 33 dialysis patient panel members identified at the T3 interviews and 26 family members or friends identified as significant others by the patient panel. This group included 12 spouses of dialysis patients, 4 friends, 3 mothers, 1 cousin, and 5 daughters and 1 son, and 45 long-term hemodialysis unit caregivers (including 8 head nurses, 14 staff nurses, 12 therapists, 3 physicians, 4 social workers, 2 machine technicians, and 2 clerks).

Procedure

Focused interviews were conducted either at hemodialysis units where patients were receiving treatment, in homes of patients' friends or family members, or in neutral settings such as a coffee shop, if so requested by a

---

*Findings at T2 showed that only 2 patients reported both serious access problems, and 2 stated severe problems with the treatment procedure itself. Five respondents mentioned social system disruptions relative to the incidence of illness or death among family or close friends, and 13 patients related serious encounters with physical illness other than chronic renal disease. The majority of patients, however, reported that these problems had been resolved at present and that their functioning on dialysis was relatively without incident.

[†]For a discussion of the utilization of Person r with ordinally scaled data see O'Brien RM: The use of Pearson's R with ordinal data, Am Sociol Rev 44:851–857, 1979.

respondent. Whenever possible and appropriate, these interviews were tape-recorded. Handwritten notes were also made during the interviews to ensure preservation of data; postinterview impressions and comments were recorded by the interviewers.

Instrumenation

Interview guides were developed for focused unstructured interviews carried out with the dialysis patient panel group, their identified significant others, and dialysis unit caregivers. These instruments were based upon previous study findings as well as the extant literature, and focused upon psychosocial variables relevant to the maintenance hemodialysis experience.

Patient Interview Guide

The patient interview guide included such foci as family relationships, support of family and friends, changes in lifestyle over time, modified or blocked life goals, quality of life, relationships with caregivers, and compliance with the therapeutic regimen.

Family Interview Guide

The instrument used to guide interviews with patients' family members or friends included such topics as impact of dialysis on family life and social activities, family support of the patient, anxieties and concerns about the dialysis regimen, modification of family goals, and difficulty of coping with ESRD and dialysis over time.

Caregiver Interview Guide

Focused guides developed for caregivers' interviews included certain topics as primary care-giving activities: degree of patient involvement, patient deaths, dealing with the long-term patient, coping with noncompliant behavior, and stresses in the dialysis unit. All interviews, however, were open-ended, respondents being asked to articulate freely their own ideas and concerns. Certain emerging themes of attitude and behavior were pursued among the study subjects as the interviewing continued.

Data Analysis

At T3 the qualitative method of data collection and analysis, as articulated by Glaser and Strauss for the discovery of grounded theory[9] was utilized. An attempt was made to cluster individual patient attitudes and behaviors into meaningful categories that might have some theoretical import in evaluating the "career" of the chronic dialysis patient. A number of emerging categories related to such topics as sick role behavior, compliance with the treatment regimen, attitudes toward dialysis, interaction with dialysis caregiver, social relationships, family role behaviors, and work activities. As these categories became more clearly defined through comparing the properties of individual patient attitudes and behaviors comprising a category, the analysis moved on to the stage of integrating the categories and delineating the construct or theory.

During the qualitative analytic process, raw data were independently assessed for categorical assignment by a medical sociologist colleague in order to provide a measure of validation for integration and reduction of core data categories.

OBSERVATION: PHASE III

The Dialysis Unit as a Social System

In order to construct as complete a picture as possible of the life career of the maintenance dialysis patient, nonparticipant observation of the sociomedical environment of the hemodialysis unit was carried out. The purpose was to attempt to identify relevant characteristics, attitudes, and behaviors of actors in the setting. Observation was done in 3 large urban outpatient dialysis units by 2 trained observers. In order to sensitize the observers to possible appropriate data sources, an observation guide was constructed. Broad categories of observational interest in the instrument included the physical environment, the technical environment, the social/interactional environment, formal activities of the actors, informal (and unplanned) activities of the actors, the language, and nonverbal communication.

Data analysis was carried out through the identification of primary themes of attitude and behavior evidenced in the observational notes. Findings were categorized through utilization of the Loomis' Model for Social System Analysis,[10] which identified such system characteristics as goals, norms, sanctioning, power, and sentiments of the actors.

INTERVENTION: PHASE IV

The Exercise of Self-Care Agency in Hemodialysis Adaptation

A clinical intervention study is presently underway to implement and evaluate a care-giving strategy: the hemodialysis education and support program. This project is directed toward promoting the concepts of autonomy, wellness, and self-care among new maintenance dialysis patients. The research is based upon previous study findings and is undergirded by the theoretical framework of self-care agency. The design is a pretest, post-test control group experimental design with repeated measures.

Criteria specify that study subjects be new dialysis patients (1–6 months on machine) and possessed of no serious physical or psychological problems other than ESRD. *Hypothesis:* chronic dialysis patients who have participated in the hemodialysis education and support program (HESP) will demonstrate more positive physical and psychosocial adaptation than patients who have not participated in the program.

Instrumentation

Physical and psychosocial adaptation is measured by several structured instruments, including Exercise of Self-Care Agency Scale, Sickness Impact Profile, Hemodialysis Regimen Compliance Scale, Inventory of Social Functioning, and Dean Alienation Scale. All study patients are pretested approximately 1 week prior to initiation of the nursing intervention strategy and post-tested 1 month, 3 months, 6 months, and 1 year following completed manipulation of the independent variable. The patients are also asked to keep a diary for the study year to record problems such as physical illness, physician visits, and hospitalizations.

Findings of the study will contribute to the body of knowledge in the area of physical and psychosocial adjustment to maintenance dialysis. Demonstration of the potential benefits of the proposed Hemodialysis Education and Support Program will also provide empirical validation for the structured intervention plan that can be utilized by practicing dialysis caregivers.

REFERENCES

1. Mead GH: Mind, Self, and Society. Chicago, University of Chicago Press, 1934
2. Duff R, Hollingshead A: Sickness and Society. New York, Harper and Row, 1968, p 3
3. Parsons T: The Social System. New York, The Free Press, 1951, p 436–437
4. Kassebaum GC, Bauman BO: Dimensions of the sick role in chronic illness. J Health Human Behavior 6:16–27, 1965
5. Durkheim E: Suicide. New York, The Free Press, 1951
6. Bridgford MH: Self Conception and Social Rehabilitation of Kidney Transplant Patients and Chronic Hemodialysis Patients. Unpublished doctoral dissertation, University of Colorado, 1970
7. Dean DG: Alienation: Its meaning and measurement. Sociol Rev 26:753–758, 1961
8. Mechanic D: Medical Sociology. New York, The Free Press, 1968
9. Glaser B, Strauss A: The Discovery of Grounded Theory. Chicago, Aldine, 1967
10. Loomis CP: Social Systems. Princeton, Van Nostrand, 1960

Index

AANNT, *see* American Association of
  Nephrology Nurses and
  Technicians
Adaptation, *see* Early adaptation; Long-
  term adaptation
Alienation, measurement of, 196
American Association of Nephrology
  Nurses and Technicians, 73
American Association of Nephrology
  Social Workers, 5
American Medical Association, 73
American Nurses Association, 74
American Society for Artificial Organs,
  74
American Society for Extra-Corporeal
  Technology, 73
Artificial kidney, 103–104, 115, 175

Behavior norms, of hemodialysis
  patient, 120–222
Break room, relationships in, 109–110
Burnout
  of caregiving staff, 85–91, 171
  defined, 87
  family stress and, 62–63, 171

Cab drivers, relationships with, 53–55
CAPD, *see* Continuous ambulatory
  peritoneal dialysis

"Career" dialysis patient, 164–165
  analysis of, 198–199
Caregiver(s), 70–100, *see also*
  Caregiving staff; Staff
  attitudes *vs.* turnover in, 119
  burnout of, 85–91, 171
  care settings and, 77–78
  characteristics of, 71–72
  as confidante, 98–100, 172
  as counselor, 97–98, 172
  and death of patient, 91–94, 124–
    127
  dialysis machine-tender as, 95–97,
    172
  early adaptation and, 131–132
  nurse as, 73–75
  patient selection dilemmas and, 123–
    127
  patient well-being and, 103
  physician as, 72–73
  preparation, experience, and
    turnover of, 78–79
  social worker as, 76–77
  stress and, 118–120
  testing of by patient, 83
  therapist/technician as, 75
  types of, 71–73, 95–98, 171–172
Caregiver interview guide, in research
  program, 198

Caregiver relationships, of hemodialysis
     patient, 29–30
Caregiving, technical aspect of, 103
Caregiving staff, *see also* Caregiver
   burnout of, 85–91, 171
   elitism of, 79–80
   preparation and experience of, 78–
      79
   support systems for, 81
   turnover of, 78–79
Caregiving strategy, evaluation of, 199
Caregiving typology, 95–100
Care settings, 77–78
Children, of hemodialysis patient, 59
Christianity, illness and, 34
Chronic ambulatory peritoneal dialysis,
      see Continuous ambulatory
      peritoneal dialysis
Chronic dialysis, treatment prototype
      in, 183
Chronic hemodialysis patient, *see also*
      Hemodialysis patient; Long-term
      dialysis patient
   career analysis of, 6–9, 198–199
   early adaptation in, 129–130
Chronic illness, patient relationships in,
      85–91, 130
Compliance behavior
   in long-term dialysis patient, 157–
      158
   sociodemographic variables in, 158–
      159
   survivors and, 159–162
Confidante, caregiver as, 98–100, 172
Continuous ambulatory peritoneal
      dialysis
   as alternative for ESRD patient, 174
   first introduction of, 3
   long-term prognosis in, 43
   physical and psychological freedom
      in, 178
   rationale for, 177–179
Continuous cyclic peritoneal dialysis,
      174
   rationale for, 179–180
Counselor, caregiver as, 97–98, 172
Courage to survive, 188–189, *see also*
      Survivor(s)

Dean Alienation Scale, 138, 196, 200
Death
   attitudes toward, 92
   caregiver's coping with, 91–95
   patient's reaction to that of fellow-
      patient, 32–34
Decreased body image, in female
      patients, 20
Dependency, 17–18
Depression, 15, 139
Despondency, 33
Dialysis, *see also* Hemodialysis;
      Maintenance dialysis
   advances in, 175
   history of in U.S., 2–4
   intermittent peritoneal, *see*
      Intermittent peritoneal
      dialysis
   peritoneal, *see* Peritoneal dialysis; *see*
      also Continuous ambulatory
      peritoneal dialysis
Dialysis caregiver, *see* Caregiver(s)
Dialysis center personnel, evaluation
      of, 135–136, *see also*
      Caregiver(s)
Dialysis machine
   ambivalence toward, 112–113
   as artificial kidney, 103–104, 115,
      175
   chair and, 105–106
   as death reminder, 32–33
   as gift or punishment, 112–113
   "going on" and "coming off" rituals
      associated with, 113–115
   language of, 115–116
   personal attachment to, 104
   physical environment and, 104–105
   social environment of, 106–107
   staff–patient interaction and, 107–
      108
   technical environment of,
      103–104
   territoriality of, 105–106
Dialysis machine-tender, as caregiver,
      95–97, 172
Dialysis patient, *see* Hemodialysis
      patient; *see also* Long-term
      dialysis patient

Dialysis treatment regimen
  evaluation of early adaptation to, 173
  family involvement with, 60–62
  long-term compliance with,
      156–157
  patient noncompliance with, 86
  routinization of, 65–66
Dialysis unit
  authority and decision making in,
      110–112
  ethical dilemmas, 123–127
  goals of, 102–103, 172
  internal and external environments
      of, 102–107
  following kidney transplant rejection,
      182–186
  language of, 115–116
  noise level of, 117–118
  as social system, 101–127, 199
  staff of, *see* Staff
  stress in, 116–120, 130
  type and size of, 77–78, 102–104
  waiting room and, 108–109
Dietary indiscretions, in long-term
      dialysis patient, 159–160

Early adaptation, 129–147
  age and, 134
  chronic illness and, 129–130
  evaluation of by dialysis center
      personnel, 135–136, 173
  socioeconomic status and, 135
  stresses and, 129–133
Early regimen compliance, 133–135
End-stage renal disease
  alternatives in, 174–182
  children's reaction to, in parents, 59
  early adaptation in, 129–147
  family interaction in, 55–59, 146
  guilt as family stressor in, 65
  kidney transplantation in, 180–182
  treatment choices for, 11, 178–182
End-stage renal disease patients, 11–44,
      see also Hemodialysis patients
End Stage Renal Disease Program, 2,
      186–187
End-stage renal failure
  acceptance in, 41

causes of, 4
  quality of life in, 39–42
ESRD, *see* End-stage renal disease
Ethical dilemmas, in dialysis unit, 123–
      127
Exercise of Self-care Agency Scale, 200

Family
  burnout of, 62–64, 171
  centripetal *vs.* centrifugal orientation
      of, 171
  concept of, 48
  defined, 48
  in dialysis treatment regimen, 60–62
  dialysis unit staff and, 62
  ESRD impact on, 55–59
  guilt feelings of, 65
  in long-term survival, 171
  overprotection by, 50
  significant others and, 49–54
  size of, 49
  social activities in, 58–59
  stresses in, 62–63
Family interview guide, in research
      program, 198
Family life, of hemodialysis patient, 22–
      23
Family member, as kidney donor, 66–
      67
Family relationships, 26–28, 48–67
Female hemodialysis patient
  decreased body image in, 20
  household role of, 56–57
"Free-lance" dialysis patient, 165–167,
      174
Friendship relationships, 28–29, *see
      also* Patient-to-patient
      relationships
Frustrations, of caregiving staff, 85–91

"Going on" and "going off" rituals,
      113–115, 172
Gore-Tex graft, 3
Guilt, as family stressor, 65

Hemodialysis, *see also* Dialysis;
      Hemodialysis patient
  care setting for, 77–78

chronic, 185–186
early adaptation to, 129–147
family reaction to, 55–59, 146
family strain in, 5, 62–63
future of, 186–187
history of, 2
home, 3, 176
in-center, 3
vs. kidney transplantation, 180–182
following kidney transplant rejection, 182–186
long-term adaptation to, 150–167
long-term view of, 170–174
patient-to-patient relationships in, 30–34
stressful nature of, 11
survivors in, 151–153, 161–167
time required per week, 5
treatment prototype for, 185–186
uremia and, 4–5
Hemodialysis adaptation, self-care agency in, 199–200, *see also* Early adaptation; Long-term adaptation
Hemodialysis Education and Support Program, 200
Hemodialysis lifestyle, accommodation to, 5, 130–140
Hemodialysis patient, 11–44, *see also* Long-term dialysis patient
accommodation to lifestyle as, 136–140
adjustments faced by, 5
age, sex, and ethical backgrounds in, 12–13
alienation in, 15–17
bargaining by, 83
"before-and-after" attitude in, 43–44
behavior norms for, 120–122
cab drivers and, 53–55
"career" type, 164–165, 174
caregiver and, 29–30, 41–42
children of, 59
as chronic illness patient, 130
courage to survive in, 188–189
death of, 91–94, 124–127
dependency in, 17–18, 84–85, 139–140

depression in, 15, 139
despondency in, 33
dialysis unit stress and, 116–120
"disenchantment and discouragement" of, 138–139
early regimen compliance of, 133–135
ethical dilemmas involving, 123–127
family attitudes toward, 50–52
family life of, 22–23
family relationships of, 26–28
family support for, 27
"free-lance" type, 165–167, 174
friends of, 28–29, 91–94
friends' death and, 32–34, 91–94
frustration in, 139
"going on" and "going off" rituals of, 113–115, 172
"good" self-portrayal by, 135
as invalid, 50
life career of, 6–9
life goals modification by, 38–39
lifestyle of, 5, 136–140
loneliness of, 16
long-term, *see* Long-term dialysis patient
long-term compliance behavior in, 157–158
machine, attitude toward, 113
machine language and, 115–116
marital relationship and, 24–25, 50–52
negative sanctioning and, 122
noise level and, 117–118
noncompliance with regimen by, 86–87
other-patient relationships and, 30–32, 91–94
parent-child relationships of, 55–57
"part-time," 165–167, 174
patient-selection dilemmas and, 123–127
personality changes in, 14–26
pity for, 50–51
psychological variables in, 152
quality of life of, 39–42
recreation for, 21–22
religion and spirituality of, 34–38

religious affiliation of, 13, 35
religious behavior of, 37–38
role reversal and, 55–56
secondary gain concept and, 144–147
self-care and self-responsibility of, 111, 140
self-perception modes in, 142–143, 173
sexual functioning in, 23–26, 138
and sickness–wellness self-perception continuum, 9, 140–147
sick role behavior of, 7, 109
significant others and, 14–15, 26–29, 49–55, 155–156
social alienation in, 15–17
social functioning of, 20–23
social–psychological functioning in, 193–195
social–psychological stress in, 5–6
spouses of, 14–15, 27, 50–51, 56, 60–64, 82–83
staff frustrations and, 85–91
staff manipulation by, 83–84
staff relationships with, 81–85, 89–91
stigma felt by, 18–20
stress and, 5–6, 116–120, 130–131
study group in, 12–14
study program for, 194–196
support systems for, 52
as survivor, 151–153, 159–167, 188–189
testing of caregiver by, 83
typology of, 163–167
ultimate goal of, 42–43
uncertain future of, 42–44
waiting-room relationships of, 108–109
work attitude of, 166
work status of, 21
in younger population, 19
Hemodialysis Regimen Compliance Scale, 200
Hemodialysis unit, *see also* Dialysis machine; Dialysis unit
in break room or lounge, 109–110

primary goal of, 172
Hemodialysis unit social worker, as caregiver, 72
HESP, *see* Hemodialysis Education and Support Program
Home hemodialysis
beginnings of, 3
choice of, 176
Household role, of female dialysis patient, 86–87
Hyperacute rejection, in kidney transplantation, 183

IBS, *see* Interactional Behavior Scale
Illness
Christianity and, 34
chronic, *see* Chronic illness
In-center hemodialysis treatment, 3
rationale for, 174–176
settings for, 77–78
Interactional behavior, 194
Interactional Behavior Scale, 195
Interactional quality, in study program, 194
Intermittent peritoneal dialysis, 3, 174, *see also* Hemodialysis
rationale for, 177
Intervention phase, in research program, 199–200
IPD, *see* Intermittent peritoneal dialysis

Kidney disease, *see also* End-stage renal disease
mortality in, 4
symptoms in, 4
Kidney donor, family member as, 66–67
Kidney transplantation
in end-stage renal disease, 180–182
rejection in, 182–186
Kidney transplant reject patient, special problems of, 184–185

Life goals, modification of, 38–39
Little Prince, The (Saint-Exupéry), 115
Long-term adaptation, *see also* Survivors
alienation scores in, 156

attitude and behaviors in, 162–167
compliance with dialysis regimen in, 156–157
evolution of, 173–174
physical and psychological changes in, 153–155
significant others in, 155–156
Long-term dialysis, typology of adaptation in, 150–167, *see also* Hemodialysis; Long-term adaptation
Long-term dialysis patient, *see also* Hemodialysis patient
attitudes and behaviors of, 163–167
CAPD and, 178
"career" type, 164–165
compliance behavior in, 157–158
dietary indiscretions in, 112, 159–160
"free-lance" type, 165–167, 174
neuropsychological evaluations of, 154
"part-time" type, 165, 174
physical changes in, 153–154
psychological changes in, 154–155
psychological variables in, 152

Machine technician, authority and decision-making role of, 110
Machine tender
as caregiver, 96–97, 172
Maintenance dialysis, *see also* Dialysis; Hemodialysis
current treatment facilities for, 1–2
long-term adaptation to, *see* Long-term adaptation
Marital relationship, 24–25, 50–52

Negative sanctioning, 122
Neuropsychological evaluations, of long-term dialysis patients, 154
New England Journal of Medicine, 174
Nurse
attitudes toward death, 91–93
burnout of, 87–88
as caregiver, 73–75, 89–90
as confidante, 98–100

coping with patient funerals, 90
ethical dilemmas of, 125–126
power and decision-making role of, 110
Nurse-patient relationships, 89–91

Observation phase, in research program, 199
Operationalization of variables, in research program, 194–196

Parent–child relationship, role reversal in, 557
"Part-time" dialysis patient, 165, 174
Patient deaths, other patients' reactions to, 32–34
Patient funerals, coping with, 90
Patient interview guide, in research program, *see also* Hemodialysis patient
Patient-staff relationship, in break room, 109–110
Patient-to-patient relationships, 30–32, 108–109
Peritoneal dialysis, 3, *see also* Continuous ambulatory peritoneal dialysis
Personality changes, 14–26
Perspiration, absence of, 18
Physical changes, in long-term dialysis patients, 153–154
Physician
authority and decision-making role of, 110
as caregiver, 72–73
ethical dilemmas of, 126
Polytetrafluoroethylene graft, 3
Potassium, dangerous levels of, 42
Power wielding, in dialysis unit, 110–112
Physiological/nutritional needs, in survivors, 160
Psychological/normalcy needs, in survivors, 160–161
Psychological variables, survival and, 152
Public Law, 92–603, 2, 186

Qualitative methods, in research
    program, 197–199
Quality of Interactional Behavior, 195
Quality of life, for hemodialysis patient,
    39–42
Quantitative method, in research
    program, 196–197

Recreation, for hemodialysis patients,
    21–22
Religion, spiritualityand, 34–38
Religious faith, patient's perception of,
    35–36
Renal failure, end-stage, *see* End-stage
    renal failure
Renal patients, adjustments faced by, 5,
    see also Hemodialysis patient
Research program, 192–200
    intervention in, 199–200
    observation phase in, 194–196
    qualitiative methods in, 197–199
    quantitative method in, 196–197
    study purpose of, 192
    study variables in, 193–194
    theoretical framework of, 193

Secondary gain, in sick role behavior,
    144–147, 196
Secondary gain scale, 145
Self-care agency, in hemodialysis
    adaptation, 199–200
Self-care dialysis programs, in Veterans
    Administration Hospital, 111–
    112
Self-perception modes, in hemodialysis
    patient, 142–143, 173
Self-responsibility, 111
Sexual functioning, 23–26
Sickness–wellness self-perception
    continuum, 7, 140–147, 173
Sick role behavior
    conceptualizing of, 194
    operationalization of, 195–196
    secondary gain and, 196
Significant others
    effective social environment and,
        193
    in long-term adaptation, 155–156

relationships with, 14–15, 26–29
support of, 49–55
Social alienation, 15–17
Social environment, of dialysis unit,
    106–107
Social functioning, 20–23, 193–194
    operationalization of, 195
Social/interactional needs, of survivors,
    161–162
Social–psychological functioning, 193–
    194
    operationalization of, 195
Social–psychological stress, 5–6, *see
    also* Stress
Social system, dialysis unit as,
    101–127
Social worker
    attitude of toward death, 93
    as caregiver, 76–77
Spouses, 14–15, 50–51, 56, *see also*
    Family; Marital relationship
    burnout of, 62–64
    role reversal in, 171
    support from, 27
Staff
    elitism of, 62, 79–80
    patient dependency on, 84–85
Staff burnout, 85–91
Staff control, 82
Staff manipulation, by patient, 83–84
Staff–patient interaction, 81–85
    dialysis machine and, 107–108
Staff–patient relationships
    in break room, 109–110
    informal, 89–91
Staff preparation and turnover, 78–79,
    119
Staff support systems, 81
Stigma, of hemodialysis patient, 18–20
Stress
    caregiver attitude toward, 118–120
    in dialysis unit, 116–120
    in family environment, 62–63
    staff turnover and, 119–120
Study phase, theoretical framework of,
    193
Study program, variables in, 193–196,
    see also Research program

Survivors, *see also* Long-term
      adaptation, Long term dialysis
      patient
courage needed by, 170, 188–189
long-term compliance behavior in,
      159–162
physiological/nutritional needs of,
      160
psychological/normalcy needs of,
      160–161
social/interactional needs of, 161–
      162
typology of, 163–167

Territoriality, in dialysis unit, 106–107
Therapist, burnout of, 88
Therapist/technician, as caregiver, 75

Uremia, in hemodialysis patient, 4–5
Urination loss, stigma of, 18

Waiting room, patient-to-patient
      interaction in, 108–109
Washington University School of
      Medicine, 2
Work status, for hemodialysis patients,
      21

a
b
3 c
4 d
5 e
6 f
7 g
8 h
9 i
8 0 j